Department of Homeland Security Bioterrorism Risk Assessment

A CALL FOR CHANGE

Committee on Methodological Improvements to the
Department of Homeland Security's
Biological Agent Risk Analysis

Board on Mathematical Sciences and Their Applications
Division on Engineering and Physical Sciences

Board on Life Sciences
Division on Earth and Life Studies

NATIONAL RESEARCH COUNCIL
OF THE NATIONAL ACADEMIES

THE NATIONAL ACADEMIES PRESS
Washington, D.C.
www.nap.edu

THE NATIONAL ACADEMIES PRESS 500 Fifth Street, N.W. Washington, DC 20001

NOTICE: The project that is the subject of this report was approved by the Governing Board of the National Research Council, whose members are drawn from the councils of the National Academy of Sciences, the National Academy of Engineering, and the Institute of Medicine. The members of the committee responsible for the report were chosen for their special competences and with regard for appropriate balance.

This study was supported by Contract No. HSHQDC-06-C-00046 between the National Academy of Sciences and the Department of Homeland Security. Any opinions, findings, conclusions, or recommendations expressed in this publication are those of the author(s) and do not necessarily reflect the views of the organizations or agencies that provided support for the project.

International Standard Book Number-13: 978-0-309-12028-9
International Standard Book Number-10: 0-309-12028-4

Copies of this report are available from the National Academies Press, 500 Fifth Street, N.W., Lockbox 285, Washington, DC 20055; (800) 624-6242 or (202) 334-3313 (in the Washington metropolitan area); Internet, http://www.nap.edu.

Printed in the United States of America

THE NATIONAL ACADEMIES

Advisers to the Nation on Science, Engineering, and Medicine

The **National Academy of Sciences** is a private, nonprofit, self-perpetuating society of distinguished scholars engaged in scientific and engineering research, dedicated to the furtherance of science and technology and to their use for the general welfare. Upon the authority of the charter granted to it by the Congress in 1863, the Academy has a mandate that requires it to advise the federal government on scientific and technical matters. Dr. Ralph J. Cicerone is president of the National Academy of Sciences.

The **National Academy of Engineering** was established in 1964, under the charter of the National Academy of Sciences, as a parallel organization of outstanding engineers. It is autonomous in its administration and in the selection of its members, sharing with the National Academy of Sciences the responsibility for advising the federal government. The National Academy of Engineering also sponsors engineering programs aimed at meeting national needs, encourages education and research, and recognizes the superior achievements of engineers. Dr. Charles M. Vest is president of the National Academy of Engineering.

The **Institute of Medicine** was established in 1970 by the National Academy of Sciences to secure the services of eminent members of appropriate professions in the examination of policy matters pertaining to the health of the public. The Institute acts under the responsibility given to the National Academy of Sciences by its congressional charter to be an adviser to the federal government and, upon its own initiative, to identify issues of medical care, research, and education. Dr. Harvey V. Fineberg is president of the Institute of Medicine.

The **National Research Council** was organized by the National Academy of Sciences in 1916 to associate the broad community of science and technology with the Academy's purposes of furthering knowledge and advising the federal government. Functioning in accordance with general policies determined by the Academy, the Council has become the principal operating agency of both the National Academy of Sciences and the National Academy of Engineering in providing services to the government, the public, and the scientific and engineering communities. The Council is administered jointly by both Academies and the Institute of Medicine. Dr. Ralph J. Cicerone and Dr. Charles M. Vest are chair and vice chair, respectively, of the National Research Council.

www.national-academies.org

COMMITTEE ON METHODOLOGICAL IMPROVEMENTS TO THE DEPARTMENT OF HOMELAND SECURITY'S BIOLOGICAL AGENT RISK ANALYSIS

GREGORY S. PARNELL, U.S. Military Academy, *Chair*
DAVID BANKS, Duke University
LUCIANA L. BORIO, University of Pittsburgh
GERALD G. BROWN, Naval Postgraduate School
L. ANTHONY COX, JR., Cox Associates
JOHN GANNON, BAE Systems
ERIC HARVILL, Pennsylvania State University
HOWARD KUNREUTHER, University of Pennsylvania
STEPHEN S. MORSE, Columbia University
MARGUERITE PAPPAIOANOU, Association of American Veterinary Medical Colleges
STEPHEN POLLOCK, University of Michigan
NOZER D. SINGPURWALLA, George Washington University
ALYSON WILSON, Los Alamos National Laboratory

Staff

NEAL GLASSMAN, Study Director, Board on Mathematical Sciences and Their Applications
KERRY A. BRENNER, Senior Program Officer, Board on Life Sciences
BARBARA WRIGHT, Administrative Assistant

For more information on the Board on Mathematical Sciences and Their Applications, see its Web site at http://www7.nationalacademies.org/bms/, write to BMSA, National Research Council, 500 Fifth Street, N.W., Washington, DC 20001, call (202) 334-2421, or send e-mail to bms@nas.edu.

Acknowledgments

This report has been reviewed in draft form by individuals chosen for their diverse perspectives and technical expertise, in accordance with procedures approved by the National Research Council's Report Review Committee. The purpose of this independent review is to provide candid and critical comments that will assist the institution in making its published report as sound as possible and to ensure that the report meets institutional standards for objectivity, evidence, and responsiveness to the study charge. The review comments and draft manuscript remain confidential to protect the integrity of the deliberative process. We wish to thank the following individuals for their review of this report:

Vicki M. Bier, University of Wisconsin, Madison,
Baruch Fischhoff, Carnegie Mellon University,
Edward H. Kaplan, Yale School of Management,
Terrance Leighton, Children's Hospital Oakland Research Institute,
Edward L. Melnick, New York University,
Tara O'Toole, Center for Biosecurity of the University of Pittsburgh Medical Center,
Harvey Ruben, University of Pennsylvania,
Ponisseril Somasundaran, Columbia University,
William Studeman, Reston, Virginia, and
Henry Willis, RAND Corporation.

Although the reviewers listed above have provided many constructive comments and suggestions, they were not asked to endorse the conclusions or recommendations, nor did they see the final draft of the report before its release. The review of this report was overseen by B. John Garrick, independent consultant. Appointed by the National Research Council, he was responsible for making certain that an independent examination of this report was carried out in accordance with institutional procedures and that all review comments were carefully considered.[1] Responsibility for the final content of this report rests entirely with the authoring committee and the institution.

The committee also acknowledges the valuable contribution of the following individuals, who provided input at the meetings on which the interim and final report are based:

Norman Coleman, National Institutes of Health,
George Famini, Department of Homeland Security,
Cyril Gay, Department of Agriculture,
Peter Highnam, Department of Health and Human Services,
Marc Lipsitch, Harvard University,
Thomas McGrann, Lawrence Livermore National Laboratory,
Darrell Morgenson, Institute for Defense Analyses,
Mark Mullen, Department of Homeland Security,
Tapan Nayak, George Washington University,
Tara O'Toole, University of Pittsburgh,
James Petro, White House Homeland Security Council,
Gregory Pompelli, Department of Agriculture,
Adam Rose, Pennsylvania State University,
Raymond Schuder, Lawrence Livermore National Laboratory,
Mark Teachman, Department of Agriculture,
Detlof von Winterfeldt, University of Southern California, and
The Staff of the Battelle Memorial Institute, Columbus, Ohio.

The committee also thanks Alan R. Washburn, U.S. Naval Postgraduate School, for his thoughtful review of and report on the Department of Homeland Security's 2006 *Bioterrorism Risk Assessment,* and Marc Lipsitch, Harvard University, for allowing his remarks to the committee to be paraphrased in this report.

[1]After the report in prepublication form was returned from a required security review conducted by the sponsor, the committee made a few revisions in the text to modify statements that might be misinterpreted.

Contents

Summary

> Armed with a single vial of a biological agent small groups of fanatics, or failing states, could gain the power to threaten great nations, threaten the world peace. America, and the entire civilized world, will face this threat for decades to come. We must confront the danger with open eyes and unbending purpose.
>
> —President George W. Bush, February 11, 2004

DEPARTMENT OF HOMELAND SECURITY'S BIOLOGICAL THREAT RISK ASSESSMENT

The Committee on Methodological Improvements to the Department of Homeland Security's Biological Agent Risk Analysis was established by the National Research Council and convened in August 2006 to review the Department of Homeland Security's (DHS's) Biological Threat Risk Assessment (BTRA) of 2006. The BTRA is a computer-based tool that has been applied by DHS to assess the risk associated with the intentional release of each of 28 biological threat agents categorized by the Centers for Disease Control and Prevention.

The threat posed by biological agents employed in a terrorist attack on the United States is arguably the most important homeland security challenge of our era. Whether natural pathogens are cultured or new variants are bioengineered, the consequence of a terrorist-induced pandemic could be millions of casualties—far more than we would expect from nuclear terrorism, chemical attacks, or conventional attacks on the infrastructure of the United States such as the attacks of September 11, 2001. Even if there were fewer casualties, additional second-order consequences (including psychological, social, and economic effects) would dramatically compound the effects. Bioengineering is no longer the exclusive purview of state sponsors of terrorism; this technology is now available to small terrorist groups and even to deranged individuals.

The executive branch recognizes this grave threat, as witnessed by the following:

- Homeland Security Presidential Directive 10 (HSPD-10): *Biodefense for the 21st Century* (The White House, 2004) calls for DHS to conduct biennial assessments of biological threats, and
- Homeland Security Presidential Directive 18 (HSPD-18): *Medical Countermeasures Against Weapons of Mass Destruction* (The White House, 2007) applies some of the basic assumptions underlying HSPD-10 to chemical, biological, radiological, and nuclear (CBRN) threats, calling for an integrated CBRN risk assessment.

DHS produced its report *Bioterrorism Risk Assessment* in 2006 (DHS, 2006). The BTRA of 2006 and the DHS (2006) report, which documents the analysis, respond directly to the requirements of HSPD-10 and of the *National Strategy for Homeland Security* (Office of Homeland Security, 2002) for DHS to assess the biological weapons threat.

This committee has been called to provide an independent, scientific peer review of the methodology that led to the BTRA of 2006 and that will be the foundation for future biennial updates. At this writing, DHS is preparing a revision of its bioterrorism risk analysis responding to HSPD-18; this analysis will presumably appear, as directed, in 2008. The committee did not have the draft of the DHS report documenting the analysis of the BTRA of 2008, but it was briefed on some of the enhancements and changed procedures that will influence the BTRA of 2008 and considered all information provided in the course of its review.

The committee has identified a number of fundamental concerns with the BTRA of 2006, ranging from mathematical and statistical mistakes that have corrupted results, to unnecessarily complicated probability models and models with fidelity far exceeding existing data, to more basic questions about how terrorist behavior should be modeled. All of these issues are covered in the body of this report.

Rather than merely criticizing what was done in the BTRA of 2006, the committee sought outside experts and collected a number of proposed alternatives that it believes would improve DHS's ability to assess potential terrorist behavior as a key element of risk-informed decision making, and it explains these alternatives in the specific context of the BTRA and the bioterrorism threat.

The committee set for itself the following gauge of success for its various deliberations and its final report: If DHS follows the committee's recommendations (drawn from the individual chapters of this report and presented as a complete set in the next section), the resulting product will more reliably assess the possible acts of terrorists, will be better documented and understood by its clients, and will be more responsive and able not only to assess risk, but to effectively inform strategic investments in risk management.

HSPD-10 states:

> Another critical element of our biodefense policy is the development of periodic assessments of the evolving biological weapons threat. First, the United States requires a continuous, formal process for conducting routine capabilities assessments to guide prioritization of our on-going investments in biodefense-related research, development, planning, and preparedness (The White House, 2004).

In accord with HSPD-10, the fundamental concerns of the committee are not only modeling or mathematical details, but the provision to homeland security policy makers of better tools to use when deciding how to invest huge sums of money to protect this nation against a grave threat.

THE CHARGE TO THE COMMITTEE

The charge to the committee for this final report is as follows:

- Recommend how the methodology can incorporate changing probability distributions that reflect how various actors (e.g., terrorists, first responders, public health community) adjust their choices over time or in different contexts;
- Recommend further improvements to the consequence analysis component of the methodology, including its models of economic effects;
- Identify any emerging methods for handling large degrees of uncertainty (e.g., fuzzy logic, possibility analysis) that merit consideration for future incorporation;
- Recommend further improvements to the transparency and usability of the methodology;
- Discuss in more detail beyond the first report [the committee's Interim Report] how the methodology could be extended to risks associated with classes of agents, including enhanced or engineered agents that have yet to be developed; and
- Discuss in more detail beyond the first report the feasibility of extending the methodology to also serve as a framework for risk analysis of chemical or radioactive threats.

In order to attend to this charge, this committee reviewed all of the detail in the BTRA of 2006, interviewed its implementers, and called on outside experts. It also received briefings from DHS on planned improvements to the BTRA of 2008. During this process, the committee recorded deficiencies and recommended improvements in the assessment.

DHS intended that the BTRA of 2006 be an "end-to-end risk assessment of the bioterrorism threat" with potential catastrophic consequences to human health and the national economy and that it "assist and guide biodefense strategic planning" (DHS, 2006, Ch. 1, p. 1) in response to the HSPD-10 directive to "conduct biennial assessments of biological threats." Guided by DHS's customers for information from the assessment, the BTRA of 2006 was designed to produce assessments in the form of risk-prioritized groups of biological threat agents. These prioritized lists could then be used to identify gaps or vulnerabilities in the U.S. biodefense posture and make recommendations for rebalancing and refining investments in the overall U.S. biodefense policy. DHS has assembled a confederation of researchers and subject-matter experts and is collaborating with national laboratories that can contribute to expanding the knowledge base of bioterrorism.

RECOMMENDATIONS

Overall Assessment

The committee met on August 28-29, 2006, with representatives of DHS in response to a DHS request for guidance on its near-term BTRA development efforts. In November 2006, in response to that request and based on the information it had received at the 2-day meeting with DHS, the committee electronically issued its Interim Report (reproduced as Appendix J in this final report). Subsequently the committee received the full DHS (2006) report documenting the analysis in the BTRA of 2006. While DHS agreed with the recommendations of the Interim Report and planned to address them, the committee did not learn of any progress up to the conclusion of its deliberations in May 2007 that would obviate those recommendations, which require sustained work.

However, the content of the DHS (2006) report and information gained at additional meetings with DHS and national experts have significantly changed the committee's overall assessment of the BTRA of 2006. The committee identified errors in mathematics, risk assessment modeling, computing, presentation, and other weaknesses in the BTRA of 2006. It recommends against using this current BTRA for bioterrorism risk assessment as presented in the BTRA of 2006 or proposed for 2008. Instead, the committee offers improvements that can significantly simplify and improve future risk assessments. The improved BTRA should be used for risk management as well as risk assessment, as intended by HSPD-10.

The committee discusses the elements of risk analyses, including risk management, and identifies the crucial differences between the use of risk analysis to assess and manage the risks of natural disasters and its use to assess and manage risks from terrorist attacks. Representing terrorist decision making exclusively as random variables, as is appropriate in the case of natural disasters, is a fundamental problem with the BTRA.

Risk Analysis Lexicon

The DHS (2006) report and DHS presentations of its content use inconsistent, imprecise technical language and do not define many key terms. Clear and consistent risk analysis definitions are essential for precise technical work and clear communication with diverse stakeholders. The committee prepared a risk analysis lexicon for its own use (included as Appendix A in this final report) with definitions and their sources. It is intended to be an example of a lexicon to be used in future DHS reports and presentations.

Recommendation: The Department of Homeland Security should use an explicit risk analysis lexicon for defining each technical term appearing in its reports and presentations.

Approach to Determining the Probabilities of Terrorist Decisions

DHS has made an important contribution by structuring a nominal bioterrorist attack and identifying the bioagents that should be assessed. The committee closely examined the assumptions and the mathematical details of the BTRA of 2006 and found that there are weaknesses in the model's conception, errors in some of the underlying mathematics and statistics, and unnecessary complexity.

The BTRA represents adversarial decisions by means of probabilities assessed by subject-matter experts. However, when dealing with an intelligent, goal-oriented, and resourceful adversary (the terrorist), the exclusive use of subjectively assessed probabilities for terrorist decisions is inappropriate. For decision problems as complex as those dealt with in the BTRA, the probability that an adversary will choose a course of action should be an *output* of analysis, not an input. Accordingly:

Recommendation: To assess the probabilities of terrorist decisions, DHS should use elicitation techniques and decision-oriented models that explicitly recognize terrorists as intelligent adversaries who observe U.S. defensive preparations and seek to maximize the achievement of their own objectives.

Simplifying the Assessment of Outcome Probabilities

Decisions, by both terrorist attacker and U.S. defender, should be outputs of a decision support model. The determination of data sources and their reliability is outside the scope of this report. However, data concerning threats, resource levels, technological facts, and so forth are inputs. Adversarial decisions can be assessed by subject-matter experts, but these assessments must be conditioned on *all* of these inputs. This is a daunting task for any subject-matter expert. Appendix G of this report contains material on alternate methods that can be used to quantify uncertainty. This report explains in detail that probability theory is suited to the task and that no alternative is needed. However, this report discusses at length weaknesses in DHS's use of probability in theory, conception, and computation in the BTRA.

Instead of directly assessing conditional probabilities for outcomes, DHS subject-matter experts are asked to assess conditional probability *distributions* over the probabilities of outcomes. This complication is shown to be unnecessary; the analysis would be unchanged if only the expected value of these distributions was used.

This simplification would significantly reduce data requirements and accelerate computation. The BTRA software implementation seems to the committee to be cumbersome and slow and requires tending by its creators to produce risk assessments. The committee advises simplification so that the BTRA can be used for responsive risk assessment and risk management.

Recommendation: The event-tree probability elicitation should be simplified by assessing probabilities instead of probability distributions for the outcomes of each event.

Regarding Normalization of Risk Assessment Results

DHS has chosen to represent "normalized" relative risk, without specifying the normalization constant. This decision has obscured the results of the analysis and made it impossible to understand the results, to reproduce any particular BTRA result, or to use independent means to assess the veracity of any result. Moreover, normalization provides insufficient information for risk assessment and risk management. Homeland security decision makers and stakeholders need to see the calculated probabilities and consequences to make risk-informed decisions. This is not to say that the committee believes that precise absolute levels of probabilities and consequences can be predicted or are needed. But risk managers and decision makers need some sense of the magnitude of the probabilities and consequences, and that is not available after normalization.

Recommendation: Normalization of BTRA risk assessment results obscures information that is essential for risk-informed decision making. BTRA results should not be normalized.

Simplification of the BTRA Event Tree

The committee finds Stage 1, Frequency of Initiation [of an attack] by Terrorist Group, of the BTRA fixed-hierarchy event-tree sequence to be a distracting embellishment. Also, the representation of potential multiple (sequential) terrorist attacks in the BTRA of 2006 is incorrect, both technically and philosophically, and adds an unnecessary layer of complexity to the analysis. The computation of the expected number of attacks is shown to be mathematically incorrect, and the (random) distribution of consequences of such repeated attacks is shown to be represented incorrectly. However, even if the mathematics were correct, the committee believes that, after the first terrorist attack, all assumptions and parameter values in the BTRA would change, so that the previous risk analysis would no longer apply. Eliminating the BTRA multiple-attack feature would significantly simplify the model.

The committee also finds that some of the stages in the BTRA characterization of the steps leading to a terrorist attack might be aggregated to the minimum number of stages necessary to calculate probabilities and consequences, making data acquisition simpler without sacrificing fidelity.

Recommendation: Two significant simplifications should be made to the BTRA of 2006 event tree:

- **DHS should eliminate Stage 1, Frequency of Initiation [of an attack] by Terrorist Group, and Stage 16, Potential for Multiple Attacks; and**
- **DHS should seek opportunities to aggregate some stages of the tree to only those essential to calculate probabilities and consequences with realistic fidelity.**

Need for Transparent, User-Friendly Decision Support System

Risk assessment, such as the BTRA, has no direct impact on risk reduction. Only effective risk management strategies can reduce risk, and there are several barriers to the effective use of information from the BTRA in decision making. These include numerous stakeholders with different responsibilities, authority, and indicators of success; disparate data and data sources; and organizational friction and compartmentalization within and among stakeholders. To support risk-informed decision making and mitigate some of these problems, DHS needs transparent and user-friendly decision support models. Accordingly, the committee makes the following three recommendations.

Recommendation: Subsequent revision of the BTRA should increase emphasis on risk management. An increased focus on risk management will allow the BTRA to better support the risk-informed decisions that homeland security stakeholders are required to make.

Recommendation: DHS should maintain a high level of transparency in risk assessment models, including a comprehensive, clear mathematical document and a complete description of the sources of all input data. The documentation should be sufficient for scientific peer review.

Recommendation: Subsequent revision of the BTRA should enable a decision support system that can be run quickly to test the implications of new assumptions and new data and provide insights to decision makers and stakeholders to support risk-informed decision making.

Rapid Assessment Strategy for New Information

The committee has highlighted the dynamic nature of the biological threat and was asked to show how the BTRA might be applied to enhanced or engineered biological agents. The committee suggests a rapid assessment tool and proposes a template that suggests how to quickly estimate the threat from emerging or suspected agents to determine whether a more detailed exigent study is necessary. It agrees that this is an important goal and makes the following recommendation.

Recommendation: The BTRA should be broad enough to encompass a variety of bioterrorism threats while allowing for changing situations and new information. DHS should develop a strategy for the rapid assessment of newly recognized and poorly characterized threats.

Existing Knowledge and the Detail in Consequence Models

The committee examined the consequence analysis of the BTRA. It finds that the susceptible, exposed, infected, and recovered (SEIR) model used to analyze the health consequences of a bioterrorist attack requires, with regard to pathogens, data that do not exist. There is scant empirical basis for pathogens that have only recently been discovered in nature and with which there is little experience. Extremely limited clinical and epidemiologic data exist about many of the pathogens in the BTRA of 2006. The granularity of detail in the SEIR models is not supported by existing data on any pathogen on the BTRA list.

Recommendation: The susceptible, exposed, infected, and recovered (SEIR) model adopted by DHS is more complex than can be supported by existing data or knowledge. DHS should make its SEIR model as simple as possible consistent with existing knowledge.

Consequences Besides Mortality and Morbidity That Need to Be Modeled

DHS is planning to include second-order economic effects in the BTRA of 2008. The committee highlights those effects, including important agricultural effects, and discusses the use of cost-benefit analysis to provide a common measure.

Recommendation: While human mortality and the magnitude and duration of morbidity should remain the primary focus of DHS bioterrorism risk analysis, DHS should incorporate other measures of societal loss, including the magnitude and duration of first- and second-order economic loss and environmental and agricultural effects.

Methods for Improved Modeling of Intelligent Adversaries

The committee attaches great importance to the realistic representation of the behavior of an intelligent adversary. BTRA probabilities are conditioned on past events and are retrospective, whereas the terrorist is prospective, constantly adjusting tactics to exploit any evident weakness in U.S. defenses.

To offer some concrete examples of how to credibly represent the behavior of an intelligent adversary, the committee presents three ways to represent adversarial decisions: (1) a "bioterrorism decision model" using off-the-shelf software; (2) a tri-level decision support model to allocate defensive investments (visible to the attacker) that represents an attacker's reasonable response to observing these preparations, and reactions to any attack with the resources made available by the defensive investments; and (3) a game-theoretic model of the adversaries that randomizes expected consequences to capture the variability of outcomes. These are not mere theoretical tools, but rather substantive suggestions drawn from extensive research and experience in the military and in the private sector. These suggestions can significantly improve the credibility and usefulness of the BTRA.

Recommendation: In addition to using event trees, DHS should explore alternative models of terrorists as intelligent adversaries who seek to maximize the achievement of their objectives.

Use of Intelligent-Adversary Risk Analysis Techniques for Other Threat Areas

The committee believes that each of its suggested extensions to realistically represent adversarial behavior is applicable to biological, chemical, and/or radioactive threats. Although distinct models may need to be developed for the analysis of each of these threats, the resulting analyses can be compared on a common consequence scale to suggest and evaluate risk management strategies that encompass all terrorist threats.

Regarding the Use of the BTRA in Its Present Form

For the reasons noted in this report's recommendations and their justifying text, the committee believes that *the BTRA in its present form should not be used to assess the risk of bioterrorism threats*. For the same reasons, the committee does not recommend trying to extend the BTRA to the qualitatively different chemical and radioactive threats.

Recommendation: The BTRA should not be used as a basis for decision making until the deficiencies noted in this report have been addressed and corrected. DHS should engage an independent, senior technical advisory panel to oversee this task. In its current form, the BTRA should not be used to assess the risk of biological, chemical, or radioactive threats.

The committee takes very seriously the bioterrorism threats and potential consequences that it has had to consider in this study. It is fully aware of the potential impact of its recommendations on the BTRA of 2008 and the stakeholders who await it. However, it believes that the failure to properly model intelligent adversaries and a continuation on the path of unnecessary complexity in computer modeling and simulations will not help the United States defend against the bioterrorist threats in the 21st century and will not meet the intent of HSPD-10. Therefore, the committee unanimously believes that an improved BTRA is needed to provide a much more credible foundation for risk-informed decision making.

REFERENCES

DHS (Department of Homeland Security). 2006. *Bioterrorism Risk Assessment*. Biological Threat Characterization Center of the National Biodefense Analysis and Countermeasures Center. Fort Detrick, Md.

Office of Homeland Security. 2002. *National Strategy for Homeland Security*. Available at www.dhs.gov/xlibrary/assets/nat_strat_hls.pdf. Accessed November 1, 2006.

The White House. 2004. Homeland Security Presidential Directive 10 [HSPD-10]: *Biodefense for the 21st Century*. Available at www.fas.org/irp/offdocs/nspd/hspd-10.html. Accessed January 16, 2008.

The White House. 2007. Homeland Security Presidential Directive 18 [HSPD-18]: *Medical Countermeasures Against Weapons of Mass Destruction*. Available at www.fas.org/irp/offdocs/nspd/hspd-18.html. Accessed January 16, 2008.

1

Introduction

Biological weapons in the possession of hostile states or terrorists pose unique and grave threats to the safety and security of the United States and our allies.

Biological weapons attacks could cause catastrophic harm. They could inflict widespread injury and result in massive casualties and economic disruption.

—Homeland Security Presidential Directive 10: *Biodefense for the 21st Century*, 2004

THIS IS THE CHALLENGE

The U.S. government has made the countering of biological weapons a top priority for well over a decade. With the international community, the United States recognizes that the biotechnology revolution, which promises a better quality of life for all people, also offers the capability for misuse. Biotechnology is powerful, relatively inexpensive, and increasingly accessible to U.S. adversaries, from nation-states, to nonstate actors including terrorists, to deranged individuals. Rapid advances in molecular biology and genomics, including the introduction of new drug-resistant agents, mean that the threat is dynamic and adaptive and that attacks could be increasingly lethal. Defending against bioterrorism may be the greatest among U.S. national security challenges.

THE THREAT IS GROWING

Today the nation is a long way from being able to meet the challenges posed by a bioterrorist attack. The United States currently has little ability to prevent or detect a biological attack, and the nation's response systems are unproven. Biological weapons are easily concealed and hard to track. Biological attacks are potentially repeatable, and attribution is extremely difficult, as was learned from the anthrax attacks in the United States in the fall of 2001. A National Intelligence Council assessment in 2004 concluded that "over the next 10 to 20 years there is a risk that advances in biotechnology will augment not only defensive measures but also offensive biological warfare (BW) agent development and allow the creation of advanced biological agents designed to target specific systems—human, animal, or crop" (National Intelligence Council, 2004, p. 36). The report states further that "as biotechnology advances become more ubiquitous, stopping the progress of offensive BW programs will become increasingly difficult" (p. 36). Before September 11, 2001 (9/11), a report by the U.S. Commission on National Security in the 21st Century (commonly known as the Hart-Rudman Commission), *New World Coming: American Security in the 21st Century*, was published; the report stated that serious threats "may consist instead of unannounced attacks by subnational groups using genetically engineered pathogens against American cities" (U.S. Commission on National Security in the 21st Century, 1999, p. 2).

Improving the U.S. capability to prevent, detect, and respond to the use of biological weapons is clearly a matter of national urgency. According to recent congressional testimony by the Director of National Intelligence, al-Qaeda and other terrorist groups continue to show interest in these weapons (Negroponte, 2007).[1]

The Commission on the Intelligence Capabilities of the United States Regarding Weapons of Mass Destruction (referred to herein as the WMD Commission) in March 2005 reaffirmed the complexity, gravity, and urgency of the threat, as well as the inadequacy of the government's response. "We are concerned," the report states, "that terrorist groups may be developing biological weapons and may be willing to use them. Even more worrisome, in the near future, the biotechnology revolution will make even more potent and sophisticated weapons available to small or relatively unsophisticated groups. In response to this mounting threat, the Intelligence Community's performance has been disappointing" (WMD Commission, 2005, p. 504). In short, the WMD Commission found that the U.S. government has been unacceptably slow to develop an effective strategic capability to prevent, detect, and respond to a biological attack.

A decade ago, experts both inside and outside government argued for a strategic, collaborative, and integrated approach to risk assessment and risk management among federal, state, and local governments, law enforcement, the military, the private sector, the media, and the medical, scientific, and academic communities (Drell et al., 1999, pp. 125-126). The

[1]A critical assessment of the intelligence community's efforts, even after 9/11, to determine al-Qaeda's biological weapons capability is contained in WMD Commission (2005).

steps taken by the federal government to develop a national strategy and the collaborative network to support it (see the next section) are still incomplete. The completion of these steps would require continuous multidisciplinary analysis and engage multiple stakeholders across functional disciplines as well as across federal, state, local, and tribal governments. The anthrax attacks in the United States in the period after 9/11 added urgency to the need for such an effort.

THE GOVERNMENT HAS TAKEN ACTION

Executive and legislative actions taken since 9/11 have sharpened the federal government's focus on bioterrorism. The Congress in November 2002 passed and the president signed the Homeland Security Act (Public Law No. 107-296), which established the Department of Homeland Security (DHS) and gave it the responsibility for developing countermeasures to biological agents. In April 2004, President Bush issued Homeland Security Presidential Directive 10 (HSPD-10): *Biodefense for the 21st Century*, which directs DHS, "in coordination with other Federal departments and agencies," to conduct assessments of the biological threat (The White House, 2004).

The first Department of Homeland Security bioterrorism risk assessment—referred to in this report as the Biological Threat Risk Assessment, or BTRA—was completed on January 31, 2006. The report documenting the analysis, *Bioterrorism Risk Assessment* (DHS, 2006) was published on October 1, 2006, by the DHS Biological Threat Characterization Center (BTCC) of the National Biodefense Analysis and Countermeasures Center (NBACC). This assessment and report satisfied the requirements of the *National Strategy for Homeland Security* (Office of Homeland Security, 2002) and of HSPD-10 for DHS to assess the biological weapons threat. DHS intended that the BTRA of 2006 be an "end-to-end risk assessment of the bioterrorism threat" with potential catastrophic consequences to human health and the national economy and that it "assist and guide biodefense strategic planning" (DHS, 2006, Ch. 1, p. 1) in response to the HSPD-10 directive to "conduct biennial assessments of biological threats." Guided by the primary customers for information from the assessment—for example, the White House Homeland Security Council, the Department of Health and Human Services, various offices of the Department of Homeland Security, the Department of Agriculture, and the Environmental Protection Agency—the BTRA of 2006 was designed to produce assessments in the form of risk-prioritized groups of biological threat agents. These prioritized lists could then be used to identify gaps or vulnerabilities in the nation's biodefense posture and to make recommendations for rebalancing and refining investments in overall U.S. biodefense policy.

National Strategy for Combating Terrorism (The White House, 2006) describes U.S. efforts against terrorism of all kinds, not just bioterrorism, and serves as guidance for the specific application of efforts against bioterrorism.

The Department of Homeland Security has made the preparation against biological weapons attacks a priority and deployed the BioWatch Program to provide early warning of an outdoor pathogen release in selected areas across the United States (Congressional Research Service, 2003). The BioWatch Program has three main elements: sampling, analysis, and response. The Environmental Protection Agency maintains the sensors that collect airborne particles. The Centers for Disease Control and Prevention coordinates analyses. Local jurisdictions are responsible for the public health response to positive findings. The Federal Bureau of Investigation is designated as the lead agency for the law enforcement response if a bioterrorism event is detected.

In January 2007, the White House issued Homeland Security Presidential Directive 18 (HSPD-18): *Medical Countermeasures Against Weapons of Mass Destruction* (The White House, 2007), which builds on HSPD-10 while "maturing" some of its basic assumptions and applying them broadly to the chemical, biological, radiological, nuclear (CBRN) challenge. Significantly, HSPD-18 mandates more incremental, integrated, and flexible policies on preparedness and response to potential weapons of mass destruction attacks. It concedes that the development and stockpiling of medical countermeasures against every possible biological threat is not feasible today, and it calls for an integrated CBRN risk assessment.

THE NATIONAL RESEARCH COUNCIL ESTABLISHED THIS COMMITTEE

At the request of the Department of Homeland Security, the National Research Council established the Committee on Methodological Improvements to the Department of Homeland Security's Biological Agent Risk Analysis to provide a review, carried out in two reports (an interim report focused on near-term improvements and the final report to recommend longer-term improvements), of the methodology described in *Bioterrorism Risk Assessment* (DHS, 2006). The interim report, prepared by the committee in 2006, is included as Appendix J of the present report.

To address its charge, the committee carried out the following activities:

- It held four 2-day meetings at the National Academies in Washington, D.C., in August and November 2006 and in January and May 2007, used for information gathering and report organization and writing;
- It heard and discussed presentations from government, academic, and medical experts;
- It received briefings on risk assessment for biological pathogens from representatives of the White House Homeland Security Council, the DHS Office of Science

and Technology, DHS's National Biodefense Analysis and Countermeasures Center (NBACC), Battelle Memorial Institute, and the Homeland Security Center for Risk and Economic Analysis of Terrorism Events;
- It reviewed DHS's *Bioterrorism Risk Assessment,* published in October 2006; and
- Committee members visited the Battelle Memorial Institute in Columbus, Ohio, for further consultations on October 2-3, 2006, because NBACC contracted with Battelle to produce a computational engine to assess the "normalized risk" of 28 pathogens as that risk relates to death, morbidity, and direct economic costs.[2] In federal fiscal year (FY) 2007, DHS directed Battelle to improve and refine its probabilistic risk assessment (PRA).

COMPLETION OF THE INTERIM REPORT

The seven tasks of the committee with respect to the interim report of December 2006 (see Appendix J) were as follows:

- To assess the adequacy of the DHS's current methodology as a foundation for the desired risk analysis capabilities;
- To identify any other risk analyses that rely on the major components of the existing methodology, probabilistic risk analysis and multi-attribute risk analysis and which could guide DHS's future developments;
- To assess the feasibility of incorporating models of second-order economic effects into the methodology during FY 2007;
- To identify better methods, if any, for handling the high degrees of uncertainty associated with the risk analyses of biological agents;
- To recommend near-term improvements to enhance the transparency of the method and its usefulness to decision-makers;
- To discuss how the methodology could be extended to risks associated with classes of agents, including enhanced or engineered agents that have yet to be developed;
- To discuss the feasibility of extending the methodology to also serve as a framework for risk analysis of chemical or radioactive threats.

In the interim report, the committee made three recommendations:

- **DHS should establish a clear statement of the long-term purposes of its bioterrorism risk analysis.**
- **DHS should improve its analysis of intelligent adversaries.**
- **DHS should increase its risk analysis methodology's emphasis on risk management.**

The interim report also commented on the technical aspects of Battelle's technique and the broader suitability of PRA. At the time it was written and under the circumstances of the writing of its interim report—that is, based solely on DHS presentations made at a single 2-day meeting and prior to committee receipt of any complete written documentation of DHS's methodology—the committee was guardedly optimistic that DHS was on the right track. As is explained more fully in Chapter 3 of the present report, when the committee was able to examine DHS's *Bioterrorism Risk Assessment* (DHS, 2006), which describes the methodology of the BTRA, it found underlying the analysis several aspects of the event-tree structure that inherently limit the ability to perform reliable risk assessment and to serve as a tool for risk managers.

The committee pointed out in its interim report that the inability to model intelligent adversaries was a major weakness in the BTRA methodology, and it recommended that DHS remedy that failing. The committee agreed that other work planned by DHS for FY 2007, notably in improving the elicitation of information from subject-matter experts and improving the modeling of consequences, was of value, and so it did not believe that a wholesale course correction was needed in FY 2007.

OVERVIEW OF THE FINAL REPORT AND OF ITS RECOMMENDED METHODOLOGICAL IMPROVEMENTS

Structure of the BTRA of 2006 Examined

As indicated above, it was only after the issuance of its interim report that the committee was provided with a copy of the DHS (2006) report documenting the BTRA methodology. The committee then gained additional information at subsequent meetings (as well as at focused visits to Battelle in Columbus, Ohio, with DHS personnel) that allowed specific examination of the technical content of the DHS (2006) report. This revised and more detailed picture assembled by the committee revealed that PRA, as used in the BTRA of 2006, is the wrong framework for modeling risks that are inherently dependent on the choices made by intelligent adversaries. The normalized risk assessments produced by such a process can be biased in ways and magnitude that cannot be determined.

In Chapter 3, the committee examines the structure of the BTRA of 2006 more closely, explains the need to model

[2] In general usage, the distinction between "direct" and "indirect" costs is not precise. "Direct" refers to costs such as those associated with closing a facility or controlling an epidemic. Other, or "indirect," costs are those that result from these actions, such as lost business or reduced productivity.

intelligent adversaries, and addresses other mathematical and structural weaknesses of the BTRA. The detailed and careful mathematical description and assessment of the BTRA described in Chapters 3 and 7, representing a major activity of the committee, were not completed in time to be included in the committee's interim report. As a result, the last recommendation in Chapter 7 of this final report represents a significant change in the overall assessment of the BTRA from that made on page 12 of the interim report (page 146 of the interim report as reprinted in Appendix J): "DHS's current methodology is adequate but incomplete."

Chapter 4 establishes the need for risk management, in part by looking at the "stakeholders," or various users of DHS's assessment information, and recommends that the DHS risk analysis be part of a decision support system. In Chapter 7 and in Appendixes D, E, and F, the committee provides three methods of doing this modeling.

Hypothetical Anthrax-Attack Scenario Employed

Thoughtfully developed, scenario-based exercises can provide unique insights of value to public- and private-sector decision makers responsible for the prevention of, preparation for, and response to bioterrorism. The notional scenario that the committee employs in this report, taken from Homeland Security Council (2004), can be used to add specificity to discussions throughout the report. This scenario, involving an aerosol anthrax attack in a highly populated U.S. city, begins with a single aerosol anthrax attack delivered by a truck using a concealed improvised spraying device in one densely populated urban city with a significant commuter workforce. Anthrax spores, delivered by aerosol, result in inhalation anthrax, which develops when the spores are inhaled into the lungs and germinate into vegetative bacteria capable of causing disease. A progressive infection follows. Attacks are made in five separate metropolitan areas in a sequential manner. Three cities are attacked initially, followed by two additional cities 2 weeks later. The crisis stresses and breaks the response capabilities of all relevant public and private institutions, rapidly leading to 328,400 exposures; 13,200 fatalities; and 13,300 other casualties. The full political, psychological, social, and economic impacts of the attack adversely affect national financial markets and consumer confidence, devastate the local and regional economy, and cause public faith in government to plummet across the country.

Lexicon of Risk Terminology Developed

This final report stresses the importance of clarity, precision, and consistency in defining risk terminology. To ensure internal consistency in its own report, the committee developed a lexicon (Appendix A) which serves as an example of the sort of clear terminology that DHS should develop, adopt, and perhaps disseminate for government-wide use. The committee employs the broad term "risk analysis" to incorporate the elements of problem formulation, risk assessment, risk communication, and risk management. The committee regards the following four principles as central to the risk analysis of the bioterrorism threat:

- *Risk analysis needs to address bioterrorism uncertainties:* Probabilistic risk assessment is a proven technique that can be used for managing the risks from bioterrorism.
- *Bioterrorism risk analysis requires access to multidisciplinary expertise:* Key disciplines include biology, epidemiology, psychology, public communications, decision analysis and risk analysis, operations research, probability, and statistics.
- *Risk analysis must be responsive to dynamic terrorism threats:* Risk analysis must take into account changing threat conditions and their resource implications over time. Intelligent adversaries will adjust their strategies and tactics to counter the U.S. ability to detect, prepare for, and respond to their attacks. Therefore, the nature of risk is a continuing evolution and will always be difficult to estimate.
- *The purpose of risk assessment is to support risk management:* Policy makers should develop risk mitigation measures that are informed by risk analysis, including assessment of social, psychological, direct, and indirect economic impacts, and should apply such measures in a manner that consciously seeks to avoid unintended consequences.

Technical and Process Improvements Recommended

This final report is intended to help DHS evaluate its progress on and to improve its methodological approach to biological agent risk assessment. The committee's charge, addressed in this report, is as follows:

- Recommend how the methodology can incorporate changing probability distributions that reflect how various actors (e.g., terrorists, first responders, public health community) adjust their choices over time or in different contexts;
- Recommend further improvements to the consequence analysis component of the methodology, including its models of economic effects;
- Identify any emerging methods for handling large degrees of uncertainty (e.g., fuzzy logic, possibility analysis) that merit consideration for future incorporation;
- Recommend further improvements to the transparency and usability of the methodology;
- Discuss in more detail beyond the first report how the methodology could be extended to risks associated with

classes of agents, including enhanced or engineered agents that have yet to be developed; and

- Discuss in more detail beyond the first report the feasibility of extending the methodology to also serve as a framework for risk analysis of chemical or radioactive threats.

In January 2006, the Office of Management and Budget (OMB) issued technical guidance for risk assessment. A report from the National Research Council (NRC, 2007) entitled *Scientific Review of the Proposed Risk Assessment Bulletin from the Office of Management and Budget* identified, in the OMB guidance, many of the same problems cited in the present report: unclear technical definitions, improper uncertainty analysis and use of expected values, and poorly conceived consequence analysis. The present report recommends technical and process improvements that are intended to make DHS risk assessment methodology more understandable, more credible, easier to communicate, and both defensible and useful at every major decision-making point in a comprehensive and effective risk management system.

In Chapter 2 the committee examines the broader context of the risk assessment methodology; in Chapter 3 it examines the implementation of the BTRA by the Battelle Memorial Institute, Columbus, Ohio; and in Chapters 3 through 7 the committee recommends improvements in the methodology. The report's 13 appendixes provide the following:

- A: A lexicon containing the technical terms used in this report;
- B: A concise mathematical description of the 2006 BTRA event tree;
- C: A numerical example illustrating the simplification of probability assessment;
- D: An alternative model for risk assessment using decision trees:
- E: An alternative model for risk assessment using mathematical optimization;
- F: An alternative model and example of risk assessment using game theory;
- G: A discussion of alternative means to quantify uncertainty;
- H: A discussion of the role of interdependencies in managing risk;
- I: An independent review of the BTRA of 2006;
- J: A reprint of the committee's interim report;
- K: The meeting agendas of the committee;
- L: Biographies of committee members; and
- M: A list of acronyms used in this report.

In the committee's view, it is imperative that the bioterrorism threat risk assessment be used to facilitate a coherent strategy of risk management against a grave and growing threat to U.S. security. The committee believes that its work will assist the federal government, as a top priority, to mature the DHS risk assessment methodology as the foundation of risk management by all the relevant stakeholders.

REFERENCES

Congressional Research Service. 2003. *The Biowatch Program: Detection of Terrorism.* Report RL32152. Available at www.fas.org/sgp/crs/terror/RL32152.html#_1_1. Accessed July 23, 2007.

DHS (Department of Homeland Security). 2006. *Bioterrorism Risk Assessment.* Biological Threat Characterization Center of the National Biodefense Analysis and Countermeasures Center. Fort Detrick, Md.

Drell, Sidney D., Abraham D. Sofaer, and George D. Wilson (eds.). 1999. *The New Terror: Facing the Threat of Biological and Chemical Weapons.* Stanford, Calif.: Hoover Institution Press.

Homeland Security Council. 2004. "Scenario 2: Biological Attack—Aerosol Anthrax," in *Planning Scenarios*. July. Available at www.globalsecurity.org/security/library/report/2004/hsc-planning-scenarios-jul04.htm#toc. Accessed November 14, 2007.

National Intelligence Council. 2004. *Mapping the Global Future: Report of the National Intelligence Council's Project Based on Consultation with Nongovernmental Experts Around the World.* Washington, D.C.: U.S. Government Printing Office.

Negroponte, John D. 2007. *Annual Threat Assessment* (unclassified for the Record). Testimony before the Senate Select Committee on Intelligence. U.S. Senate. Washington, D.C. January 11.

NRC (National Research Council). 2007. *Scientific Review of the Proposed Risk Assessment Bulletin from the Office of Management and Budget.* Washington, D.C.: The National Academies Press.

Office of Homeland Security. 2002. *National Strategy for Homeland Security.* Available at www.dhs.gov/xlibrary/assets/nat_strat_hls.pdf. Accessed November 1, 2006.

U.S. Commission on National Security in the 21st Century. 1999. *New World Coming: American Security in the 21st Century.* Washington, D.C.: U.S. Government Printing Office.

The White House. 2004. Homeland Security Presidential Directive 10 [HSPD-10]: *Biodefense for the 21st Century.* Available at www.fas.org/irp/offdocs/nspd/hspd-10.html. Accessed January 16, 2008.

The White House. 2006. *National Strategy for Combating Terrorism.* Available at www.state.gov/s/ct/rls/wh/71803.htm#overview. Accessed July 23, 2007.

The White House. 2007. Homeland Security Presidential Directive 18 [HSPD-18]: *Medical Countermeasures Against Weapons of Mass Destruction.* Available at www.fas.org/irp/offdocs/nspd/hspd-18.html. Accessed January 16, 2008.

WMD Commission (Commission on the Intelligence Capabilities of the United States Regarding Weapons of Mass Destruction). 2005. *The Report on the Intelligence Capabilities of the United States Regarding Weapons of Mass Destruction.* Available at www.wmd.gov/report. Accessed January 16, 2008.

2

The Critical Contribution of Risk Analysis to Risk Management and Reduction of Bioterrorism Risk

> Risk management must guide our decision making as we examine how we can best organize to prevent, respond, and recover from an attack.
>
> —Department of Homeland Security Secretary Michael Chertoff at Homeland Security Policy Institute, March 16, 2005

Homeland Security Presidential Directive 10 (HSPD-10): *Biodefense for the 21st Century* (The White House, 2004) cites two applications for which a bioterrorism risk assessment is needed: the identification of gaps or vulnerabilities in the U.S. biodefense posture and the rebalancing and refining of investment in U.S. biodefense policy. The list of "stakeholders," or primary public-sector customers, as identified by the Department of Homeland Security (DHS), is presented in Chapter 4. Although the committee does not know the uses to which these stakeholders will apply the Biological Threat Risk Assessments (BTRAs) of DHS, it is confident that these uses and the two explicitly mentioned in HSPD-10 will require the use of the BTRA as the basis of a risk analysis system. This chapter examines the components of such a system, especially as they relate to health risk analysis.

RISK ANALYSIS IS THE DISCIPLINE THAT THE DEPARTMENT OF HOMELAND SECURITY SHOULD USE

The risk analysis framework consists of five elements:

- Problem formulation,
- Risk assessment,
- Risk perception,
- Risk communication, and
- Risk management.[1]

Risk analysis offers (1) a framework for applying scientific knowledge and the data to examine risk management decision making when the consequences of alternative decisions are uncertain and (2) a systematic method of revising decisions in the light of new information or events. The hazards to be analyzed (e.g., physical, chemical, nuclear, radiological, and biological agents) may result from natural events (e.g., earthquakes and hurricanes), technological events (e.g., chemical accidents), and human activity (e.g., the design and operation of engineered systems or an attack by a terrorist).

In *Bioterrorism Risk Assessment*, the DHS (2006) report describing the methodology of the BTRA of 2006, DHS used only two of these elements, problem formulation and risk assessment, as described in Chapter 3 of the present report. However, the committee believes that all five steps listed above should be unified and taken with the ultimate goal of effective risk management.

Problem Formulation

To undertake any systemic risk analysis, it is necessary to clarify the problem being studied, the key stakeholders, their relationship to one another and to the problem being solved, and their values and goals (Keeney, 1992). Stakeholders may have different objectives, depending on the potential type of attack being considered: for example, for some, prevention may be the primary concern; for others, response and mitigation may be primary. Without a clear understanding of stakeholder objectives with respect to alternative terrorist tactics, risk management strategies may be developed that are unlikely to be implemented. In the context of the bioterrorism problem, the key interested parties are the relevant public-sector agencies concerned with this risk, the terrorists (who would like to discover U.S. assessments and policies), those who will be directly and indirectly attacked by the terrorists, those adversely affected economically and physically (through adverse health effects), and the taxpayer, who will have to pay for the risk management and some of the losses.

In order to make the best choices, public and private decision makers may require inputs from biologists, public health care professionals, decision analysts, risk analysts, economists, political scientists, policy analysts, psychologists, sociologists, statisticians, and related professionals. Since

[1] For more details on the risk analysis framework, see Kunreuther (2002).

by the committee's definition almost everyone in the U.S. population is a stakeholder in BTRA information, it is important to develop strategies to reconcile differences among subpopulations. These subpopulations will perceive risk on the basis of their own goals and objectives. Techniques such as value-tree analysis (von Winterfeldt, 1987) may be useful in bringing out and reconciling these differences.

Risk Assessment

Risk assessment is the process of identifying hazards and targets and quantifying the risks that the hazards pose (magnitude, spatial scale, duration, and intensity) and the associated probabilities, including the uncertainties surrounding these estimates.[2] The primary goal of risk assessment is to produce information to improve risk management decisions by identifying and quantifying cause-and-effect relationships between alternative risk management decisions and their consequences and by identifying decisions that may increase the probabilities of preferred outcomes. Risk assessment may include a description of the cause-and-effect links between different hazards, and the nature of the interdependencies, vulnerabilities, and consequences.

Once the problem has been formulated, risk assessment begins with *hazard identification:* the process of specifying the scope of the assessment and summarizing the available empirical evidence showing that a specific "hazard" (such as exposure to a specific pathogen in a specific environment) causes specified adverse health effects. Hazard identification can serve the following purposes:

- Rapid screening of potential hazards by identifying whether available data support the hypothesized relationship between the hazard and specific health effects, possibly using formal statistical methods of causal analysis (Shipley, 2000);
- Identification of causal relationships between identified hazards and specific adverse human health effects; and
- Identification of risk factors, behaviors, and exposure conditions that increase risks to specific exposed populations (e.g., the old, the young).

Studies to identify specific hazards, their probability of occurrence, and the probability of occurrence of their associated consequences are a part of risk assessment. In these studies, experts can provide insight into terrorists' values and objectives—along with their assessments of associated risks—but the experts need to take special care not to filter these estimates through their own values.

Health risk assessments are specializations of the methods described above. They typically use explicit analytic models (e.g., statistical models, probabilistic simulation) of causal relationships between actions and their probable health effects. Exposure models describe the transport and distribution of hazardous materials through different media and pathways (e.g., air, foods, drinking water) leading from their source(s) to members of the exposed population. Because different exposures lead to different health outcomes, a successful exposure assessment should describe the frequency distribution of exposures of different parts of the population.

Dose-response models ideally quantify the conditional probability of illness caused by each level of exposure as well as the degree of uncertainty surrounding these estimates. For some biological agents, it may be necessary to fit separate dose-response models to "normal" and "susceptible" subpopulations at risk and to account for interindividual variability in dose-response relations. In general, risk assessment requires a description of the *severities* as well as the *frequencies* of adverse health outcomes caused by exposures and the potential value of gathering additional information to reduce the uncertainty surrounding these risk estimates.

One useful graphical way to capture the extent of expert knowledge about a particular risk is to construct an exceedance-probability (EP) curve. An EP curve specifies the probability that a certain level of losses will be exceeded. The losses can be measured in terms of dollars of damage, fatalities, illness, or some other unit of analysis. If one views the loss as a random variable, the EP is simply the complementary cumulative distribution of the loss.

For example, suppose one were interested in constructing an EP curve for direct dollar losses from the first bioterrorism attack described in the aerosol anthrax scenario employed in this report (see Chapter 1). Event trees and fault trees,[3] used as part of probabilistic risk assessments, would identify the set of conditions and subsequent events that could produce a given dollar loss, determine the resulting probabilities of exceeding losses of different magnitudes, and combine the results. Based on these estimates, the mean EP curve, depicted in Figure 2.1, could be constructed. Suppose that one focuses on a specific loss, L_i. One can see from Figure 2.1 that the likelihood that losses will exceed L_i is given by p_i. The x axis measures the loss in dollars and the y axis depicts the probability that losses will exceed a particular level.[4]

It is much easier to construct an EP curve for natural disasters and chemical accidents than for bioterrorist activities. But even for those more predictable accidents or disasters, there may be considerable uncertainty regarding the occurrence of certain risks and the resulting damage. Providing information on the range of this uncertainty asso-

[2]See Haimes (1998) for a comprehensive summary of recent work in risk assessment.

[3]See the lexicon in Appendix A for definitions of *event tree* and *fault-tree analysis*.

[4]A detailed discussion of how one constructs an EP curve and incorporates elements of uncertainty on these estimates appears in Grossi and Kunreuther (2005, Chapters 2 and 4).

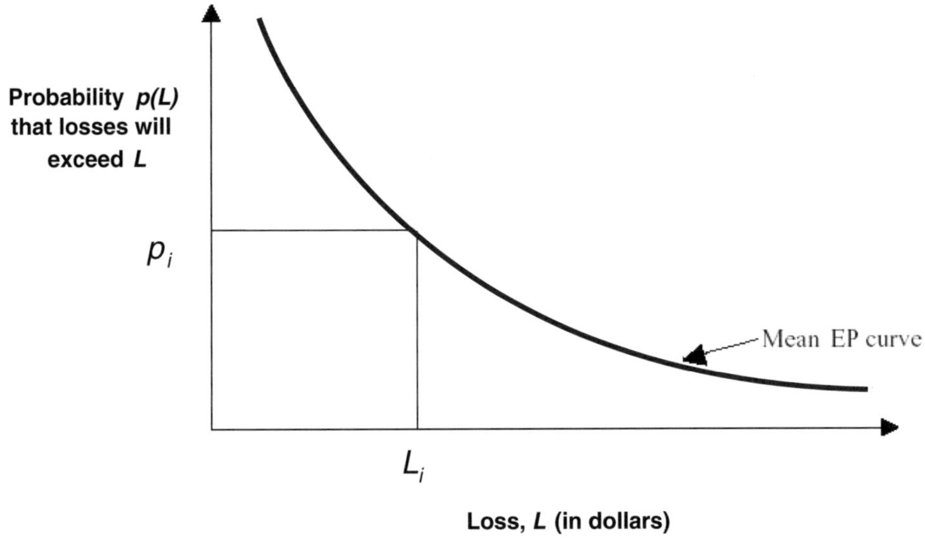

FIGURE 2.1 Example of a mean exceedance-probability (EP) curve.

ciated with risk assessments should increase the credibility of the expert estimates of these numbers.

The model used for the DHS BTRA of 2006, fully described in Chapter 3 of this report, was used to determine the relative risk of the terrorist use of each of 28 specific pathogens, identified in other sources.

Risk Perception

Risk perception is concerned with the psychological factors, including emotional factors, that have been shown to have an enormous impact on behavior (Slovic, 2000). Risk perceptions can be influenced by personal knowledge, experience, and beliefs, and they can be affected by an individual's changing recognition of the threat, the vulnerabilities, and/or the consequences. Risk perception may be influenced by new information about hazards, risk assessments, risk policies, and risk management decisions.

In a set of pathbreaking studies begun in the 1970s, psychologists began measuring laypeople's concerns about different types of risks. These studies showed that those hazards for which a person had little knowledge and which were also highly dreaded were perceived as being the most "risky" (e.g., most probable). For some technologies such as nuclear power and activities such as storing radioactive waste, there was a wide disparity between the general citizenry's view and the experts' view of the risk—that is, of both the hazards and their associated probabilities. The finding that laypeople and the scientific community see the world differently also raised a set of questions as to the nature of the decision-making process for dealing with risks.

For some time those in the scientific community felt that it was appropriate to ignore the public's perception of the risk if it differed significantly from their own estimates. It is now known that the public did not believe the experts' assessments because those assessments were not communicated well, the assumptions on which they were based were not stated well, and there was little understanding by the public of the reasons for disagreement among the experts. In recent years, there has been increased sympathy for including the psychological and emotional factors involved in perception of risk as part of risk assessment.

Recent studies have confirmed this view of how the public perceives risk by showing that the public will assiduously avoid certain activities because they are perceived to be unduly dangerous. More specifically, there is a stigma associated with technologies, places, and products if the public perceives them to be hazardous (Flynn et al., 2001) even though in many of these cases the scientific evidence suggests that there is little to be concerned about. Stimulated by media reporting, the public's perception of the risk is often amplified in ways that are difficult to explain solely by a technical risk assessment (Kasperson et al., 2001).

The problems associated with risk perception are compounded because of the difficulty individuals have in making a decision requiring the interpretation of very low probabilities. In fact, there is empirical evidence that people may not even want data on the probability of an event's occurring (Huber and Wider, 1997). There is now a large body of evidence that individuals' risk perceptions are affected by judgmental biases. The *availability heuristic* is one of the most relevant biases for dealing with extreme events: here people estimate the probability of an event by the ease with which they can imagine or recall past instances (Tversky and Kahneman, 1973). In cases where the information on an event is salient, so that individuals fail to take into account the base rate, there will be a tendency by many to overestimate the probability of the event's occurring. Following the terrorist attacks of September 11, 2001 (9/11), many people refused to fly because they perceived a high probability of being hijacked. This was true even though it could be argued that the probability of being hijacked was extremely low, given the increased vigilance and added protection by the federal government.

There is also a growing body of evidence that emotions play an important role in an individual's decision processes. Such behavior is not irrational. Rather than basing one's choices simply on the probability and consequences of different events, as normative models of decision making suggest, individuals are also influenced by emotional factors such as fear, worry, and love (Finucane et al., 2000; Loewenstein et al., 2001).

Risk Communication

The importance of risk communication in the overall risk management process is emphasized in Homeland Security Presidential Directive 10 (The White House, 2004):

> A critical adjunct capability to mass casualty care is effective risk communication. Timely communications with the general public and the medical and public health communities can significantly influence the success of response efforts, including health- and life-sustaining interventions.

Risk communication is used by risk analysts, decision makers, policy makers, and even intelligent adversaries to provide data, information, and knowledge designed to change or to shape the risk perceptions of individuals and organizations and to cause them to assess the risk in a different way than they otherwise might. Well-designed risk communication facilitates the effective participation and interaction of technical experts, stakeholders, and decision makers in risk management decision processes and deliberations. Risk communication is also used to present the results of risk analyses to stakeholders, decision makers, participants, and other audiences. Communication and deliberation drive much of the risk management decision process in many risk management applications and are essential for successful outcomes. The relationship of risk communication to risk management is examined in the National Research Council (NRC) report entitled *Understanding Risk,* which states: "the process (of risk characterization) must have an appropriately diverse participation or representation of the spectrum of interested and affected parties, of decision makers, and of specialists in risk analysis, at each step" (NRC, 1996, p. 3).

The most common goals for risk communication programs are these:

- *To provide information to individuals and groups* about risks so that they can make better-informed decisions or seek more information;
- *To influence people* to change their behaviors, their attitudes, and beliefs about hazards and their acceptance of risk management decisions and policy recommendations;
- *To involve affected parties* in the decision process; and
- *To facilitate their participation* in conflict-resolution, consensus building, and collective decision making about risk management.

The field of risk communication provides guidelines for the accomplishment of these goals, derived mainly from experience, analysis of survey data, and experiments, and for sharing risk information among stakeholders and decision makers.

As noted above, a number of studies have shown that people have difficulty processing data regarding low-probability events. This raises the problem of effectively communicating information on risk to the public, especially information involving very low or high probabilities—an important component in any risk communication strategy for dealing with the bioterrorist threat. The use of EP curves such as that shown in Figure 2.1 can indicate the uncertainties surrounding a particular risk. However, as pointed out above, laypeople are not likely to process these data in the formulaic manner that scientists and engineers might. Risk communication approaches must recognize the difficulties that individuals have in collecting and analyzing data from experts, particularly with respect to low-probability events.

The format and presentation of risk information and the framing of associated questions or surveys can greatly affect the manner in which recipients respond to, assimilate, and act on the information. For example, in medical decisions, people are more likely to elect a medical procedure when it is described as "99 percent safe" than when it is described as having "a 1 percent chance of complications" (Gurm and Litaker, 2000). Presenting relative risks rather than absolute risks and using loss framing instead of gain framing make it more likely that patients will adopt screening procedures. In presenting economic risks, the language used may trigger speculations about the presenter's motives and undermine his or her credibility with the target audience (MacGregor et al., 1999). Understanding such effects can help in preparing the presentation of factual information in ways that are likely to elicit desired responses.

A striking insight from the framing literature is that there may be no *neutral* way to present risk information. Any presentation carries with it potential presentation and framing effects and biases that may affect the recipients' attention, interpretation, and actions. Presenting the same information in different ways and emphasizing fact-rich displays (e.g., cumulative risk profiles) that are not strongly associated with known presentation biases may come as close as possible to providing the information needed for rational decision making without biasing the decision. However, such displays may be difficult to understand, as they may lack the brevity and focus that are most effective in an action-oriented presentation.

The challenge of biological agent risk analysis is daunting because it requires inputs from multiple disciplines and, if properly integrated into risk management, will engage a vast network of stakeholders across every level of government, the private sector, the medical community, and the media. Progress toward this goal will require that the diverse population of stakeholders share a common language and terminology with respect to concepts of risk analysis. This concept has yet to be translated into reality.

Precise terminology has a special urgency in the case of biological agent risk analysis. As is always the case in science, the absence of a precise definition of terms frustrates the effort to improve methodologies because experts may use the same words or phrases differently. For example, the word "risk" may be interpreted in very different ways by different individuals.

The committee stresses the importance of terminology. Because the BTRA is meant to provide a basis for critical planning and decision making, some of it very costly and with its own risks, imprecision in terminology can have serious consequences. In the briefings that the committee received, there was ambiguous, conflicting, and incorrect use of some technical terms. The committee has made an effort to provide authoritative definitions of all of the relevant terms used in this report and includes the lexicon that it developed as Appendix A. This can serve as a model for a DHS lexicon.

Recommendation: The Department of Homeland Security should use an explicit risk analysis lexicon for defining each technical term appearing in its reports and presentations.

Risk Management

Risk management is the process of constructing, evaluating, implementing, monitoring, and revising strategies for reducing (or distributing) losses from future hazards and dealing with the recovery process should a hazard occur. Risk management takes scientific information obtained from risk assessment and factors influencing risk perception as inputs, along with value judgments and with policy goals and constraints, and proposes alternative strategies for reducing losses from future hazards and dealing with the recovery process should a disaster occur. Risk management strategies include a combination of options such as the provision of information (i.e., risk communication); the offering of economic incentives (e.g., subsidies, fines); prevention or avoidance (e.g., by reducing exposures); the mitigation of consequences (e.g., by appropriate clinical screening, diagnosis, and treatment procedures); and/or the transfer of risk (e.g., insurance and compensation). As with the other elements of risk analysis, it is important to identify the key stakeholders and their values and goals as well as their short- and long-term priorities. How do they perceive the risks, and what do they need from the risk assessment in order to make better resource allocation decisions?

In combination with risk communication strategies, one can employ economic incentives to encourage individuals to take protective measures against the bioterrorism threat. Fines coupled with specific regulations or standards can be used to encourage the adoption of protective measures, although there needs to be a sufficiently high probability that any negligent individual or firm will get caught. Otherwise the person or manager is likely to respond to incentives different from those intended (i.e., ignore the regulation). If the probability is low enough and/or the fine is small enough, a person may decide that it may pay in the long run not to take protective action. The behavior in such cases is similar to the decision not to put a quarter in a parking meter because one figures that there is a small chance of getting a ticket and in any case the ticket doesn't cost much.

Risk management strategies can be evaluated by undertaking cost-benefit analyses to determine the trade-off between the reduction of risk and the costs of undertaking such measures. In evaluating a risk management strategy, one needs to be concerned with the way that resources are allocated (i.e., efficiency considerations) as well as the impact of these measures on different stakeholders (i.e., distribution or equity considerations).

A successful risk analysis shows the estimated changes in the frequencies and magnitudes of adverse consequences resulting from different risk management decision options. Risk analysis uses probability distributions, confidence intervals, and other displays to show the uncertainties about the human health consequences of different decisions. It identifies a subset of decision options leading to preferred probability distributions of health risks and other outcomes.

The outputs of a health risk analysis should allow a risk manager to answer the following questions for each risk management decision alternative being evaluated or compared:

- *What change in human health risk would result from each risk management intervention?* If the risk management option or action being assessed is implemented, how will the adverse human health effects (e.g., expected numbers of mild, moderate, severe, and fatal illnesses per year; expected numbers of illness-days, duration, and latency) change, both in the entire population and in subpopulations with distinct risks?
- *How certain is the change in human health risk that would be caused by each risk management action?* Instead of a single value, that is, a point estimate of risk,

uncertain risks are characterized by intervals or probability distributions indicating how closely the change in human health risk caused by a proposed risk management intervention can be predicted. Might management action cause further damage, such as from unforeseen effects of large-scale inoculation or the administration of antidotes? There are several technical options for expressing uncertainty around point estimates (e.g., plausible upper and lower bounds, confidence limits, coefficients of variation).

- *What are the key drivers of hazards and uncertainties for each option?* The analysis should make clear to the planner the main reasons why the estimated risk from each decision option is as high or low as it is. Are the results driven mainly by predicted exposure levels, by the responses of sensitive subpopulations, by genetic or epidemiological data that establish tight constraints on the plausible values, or by other factors? Sensitivity analyses plotting the change in estimated risk as input assumptions and estimates vary within plausible ranges (e.g., within a few standard deviations of their median or mean values) and can help to identify the combinations and range of input values that drive the main conclusions.

TERRORIST THREATS DIFFER FROM NATURAL HAZARDS AND FROM OTHER HUMANLY MADE HAZARDS

A special challenge in developing risk assessments for a terrorist attack involves human action and reaction. Although terrorist activities and natural disasters can both be characterized as extreme events, there are crucial differences between them,[5] in areas including the following: the availability or lack of historical data, dynamic uncertainty, shifting of attention to unprotected targets, the existence of negative externalities, and governmental influence on the risk. These characteristics are discussed below and summarized in Tables 2.1-2.3.

Large historical databases on losses from natural hazards are available in the public domain. These data have been utilized by modeling firms in conjunction with estimates by scientists and engineers on the probability and consequences of future disasters in specific locations. In contrast, data on terrorist groups' activities and current threats are normally kept secret for national security reasons. Moreover, while some time-series data on terrorist acts over the past years are in the public domain, they may not reflect the changing expectations of planned activities of terrorist groups today.

Because terrorists are likely to design their strategies as a function of their own resources and their knowledge of the vulnerability of their specific targets, the nature of the risk is continuously evolving. The probability and consequences of a terrorist attack are determined by a mix of actions and counteractions developed by a range of involved parties and changing over time. This leads to what is called dynamic uncertainty (Michel-Kerjan, 2003). In contrast, actions can be taken to reduce damage from future natural disasters with the knowledge that the probability associated with the hazard will not be affected by the adoption of these protective measures. For instance, the probability of an earthquake of a given intensity in a specific location will not change if property owners design more quake-resistant structures.

In addition, there are issues of interdependent security that need to be considered when predicting or planning involves the actions of each individual at risk from a bioterrorist attack (Heal and Kunreuther, 2006). This interdependence, as well as issues of perception and communication, was recognized in an earlier NRC report, *Terrorism and the Chemical Infrastructure* (NRC, 2006). Even if an individual or firm has taken protective actions, there is still some chance that that entity can be contaminated or infected by others who have not undertaken similar measures and hence are at risk. For example, if a person has been vaccinated or taken preventive medicine against a disease, he or she may still contract the illness from others if the vaccine or medicine is not 100 percent effective. Even if modifications to a single unit of an organization can reduce the chance of a bioterrorist attack to its own operations, that chance can still be adversely affected by a second unit that did not undertake similar protective measures. In these cases, where there are complementarities or positive externalities created by an individual taking protective measures, there is more incentive for one unit to invest in protective measures if the other units have taken similar actions. In fact, investing in security is most effective if all elements of the system obtain protection; weak links may lead to suboptimal behavior by everyone (Heal and Kunreuther, 2006; Bier, 2007).

Information sharing about risk due to terrorism is clearly different from information sharing about risk due to natural hazards. In the latter case, new scientific studies normally are common knowledge, so insurers and the individuals and businesses at risk, as well as public-sector agencies, all have access to these findings. However, information on terrorist groups' activities, possible attacks, or current threats is kept secret by government agencies for national security reasons.

There are also more fundamental differences between the catastrophic modeling of natural hazards and the modeling of megaterrorism. The issue of effectively modeling the actions of intelligent adversaries by other than probabilistic estimates is central to this report and is addressed more fully in the remainder of the report. International terrorism is a matter of national security as well as foreign policy. The government can influence the level of risk of future attacks through appropriate counterterrorism policies and international cooperation as well as through adequate crisis

[5]For more details on these differences, see Parnell et al. (2005) and Golany et al. (2007).

TABLE 2.1 Natural Hazards Versus Terrorism Risks: Comparison of Key Characteristics

Characteristic	Natural Hazards	Terrorist Attacks
Historical data	*Some historical data:* A record exists of extreme events that have already occurred.	*Very limited historical data:* Events of September 11, 2001, were the first terrorist attacks worldwide with such a huge concentration of victims and insured damages.
Risk of occurrence	*Reasonably well defined:* Well-developed models exist for estimating risks based on historical data and expert estimates.	*Considerable ambiguity:* Terrorists can purposefully adapt their strategies depending on their knowledge of a target's vulnerabilities.
Geographic risk	*Specific areas at risk:* Areas such as California for earthquakes or Florida for hurricanes are well known for being at risk.	*All areas at risk:* Although some cities may be considered riskier than others, terrorists may attack anywhere.
Information	*Information sharing:* New scientific knowledge on natural hazards can be shared with all stakeholders.	*Asymmetry of information:* Government may keep new information secret for national security reasons.
Event type	*Natural event:* No one can influence the occurrence of extreme natural events.	*Terrorist event:* Governments can influence terrorism through foreign policy, security measures, or international cooperation.
Preparedness and prevention	*Measures known:* Investments can be made in well-known mitigation measures.	*Possible unforeseen events:* Weapons and weapon configurations are numerous, and there can be substitution in terrorist activity.
Catastrophe modeling	*Well developed:* Developed in late 1980s and early 1990s.	*Development needed:* First models developed in 2002.

TABLE 2.2 Natural Occurrence of Anthrax Versus Its Use by Terrorists: Comparison of Key Characteristics

Characteristic	Natural Occurrence of Anthrax	Use of Anthrax by Terrorists
Historical data	*Some historical data:* Good understanding exists of the modes of transmission and containment.	*Limited historical data:* Limited historical and experimental data exist. There are no data corresponding to a dispersed nationwide attack.
Risk of occurrence	*Well understood:* Risk is well understood.	*Considerable ambiguity:* There is a wide range of possible attacks using existing or unknown strains.
Geographic risk	*Specific areas at risk:* Good scientific understanding of the relationship between geography and risk of disease exists.	*All areas at risk:* Terrorists may attack anywhere with the possibility of wide geographic dispersion designed to maximize exposure. Governments can influence local risk through security measures or international cooperation.
Information	*Information sharing:* New scientific knowledge can be shared with all stakeholders.	*Asymmetry of information:* Government may keep new information secret for national security reasons.
Event type	*Natural event:* Most natural events will not be extreme, but localized.	*Terrorist event:* Terrorists will seek to maximize their objectives.
Preparedness and prevention	*Measures known:* Investments can be made in well-known mitigation measures.	*Possible unforeseen events:* Terrorists will attempt to obviate preparations, for example creating a strain resistant to the stockpiled antibiotic.

management to limit the consequences should an attack occur. Some decisions made by a government as part of its foreign policy can also affect the will of terrorist groups to attack the country or its interests abroad (Lapan and Sandler, 1988; Lee, 1988; Pillar, 2001).

A government can also devote part of its budget to the development of specific measures on national soil to protect the country. The creation of the U.S. Department of Homeland Security in 2002 confirms the importance of this role in managing the terrorist risk. In that sense, terrorism risk is partly under the government's control, and it will change depending on at least two complementary strategies by the defenders: the first entails protective measures that could be adopted by those at risk; the second consists of actions taken

TABLE 2.3 Natural Occurrence of Smallpox Versus Its Use by Terrorists: Comparison of Key Characteristics

Characteristic	Natural Occurrence of Smallpox	Use of Smallpox by Terrorists
Historical data	*Known to be noncatastrophic:* There is wide variation in the impact of smallpox between developed and developing regions. As a natural disease, smallpox is not catastrophic to U.S. interests, although it could have significant mortality and economic consequences.	*No historical data:* There is no prior experience with smallpox dispersed as a modern, large-scale attack. Fatalities could be in the hundreds or perhaps even low thousands (but since the vaccine can be usefully administered up to 7 days after exposure, early detection of an attack would be invaluable).
Risk of occurrence	*No risk:* Essentially zero risk of occurrence exists.	*Some uncertainty:* The creation or acquisition of smallpox is well within the technical reach of a determined and well-resourced terrorist, but it is not clear that such a terrorist would pursue smallpox over a radiological device or investment in conventional weapons.
Geographic risk	*Limited risk:* There is almost zero risk of occurrence, except possibly in Eritrea.	*Containment difficult:* A successful aerosol dispersion would require sophisticated technology and could fail owing to malfunction, weather conditions, or other factors. However, if an attack was carefully planned, containing the disease to the area of attack would be very difficult.
Information	*All in public domain:* Essentially all information on naturally occurring smallpox is in the public domain.	*Asymmetry of information:* If some group has created a weaponized version of smallpox, that would be a closely held secret. Similarly, technology for dispersion would be secret. If counterintelligence discovered that a terrorist group was preparing a smallpox attack, the decision on how to use that information would depend on the reliability and completeness of the information.
Event type	*Natural event:* Historically, most natural outbreaks were not extreme and "burned out" within a few months (much more rapidly in developed regions).	*Event unlikely:* The very public preparation in the United States against the possibility of a smallpox attack has probably changed the climate for terrorist thinking. Given that this nation has now stockpiled significant quantities of vaccine, and given that smallpox is slow to progress and easy to diagnose, it seems unlikely that a rational terrorist would choose this attack over comparably difficult but more consequential alternatives.
Preparedness and prevention	*Smallpox eradicated:* The eradication of natural smallpox was a great public health success.	*Vaccination possible:* Swift vaccination could protect all of the U.S. population except those affected in the first wave of an attack.

by the government to enhance the general security and to reduce the probability that attacks will occur. Hence protection from terrorism is a mixed private-public good.

Table 2.1 summarizes the distinctions between risks from natural hazards and those from a terrorist attack. Tables 2.2 and 2.3 particularize these distinctions to apply to anthrax, as in the hypothetical scenario used for this report, and smallpox.

REFERENCES

Bier, V. 2007. "Choosing What to Protect." *Risk Analysis* 27(June):607-620.

DHS (Department of Homeland Security). 2006. *Bioterrorism Risk Assessment.* Biological Threat Characterization Center of the National Biodefense Analysis and Countermeasures Center. Fort Detrick, Md.

Finucane, M.L., A. Alhakami, P. Slovic, and S.M. Johnson. 2000. "The Affect Heuristic in Judgments of Risks and Benefits. *Journal of Behavioral Decision Making* 13(1):1-17.

Flynn, J., P. Slovic, and H. Kunreuther (eds.). 2001. *Risk Media and Stigma.* London, U.K.: Earthscan.

Golany, B., E.H. Kaplan, A. Marmur, and U.G. Rothblum. 2007. "Nature Plays with Dice—Terrorists Do Not: Allocating Resources to Counter Strategic Versus Probabilistic Risks." *European Journal of Operational Research.* In press.

Grossi, P., and H. Kunreuther. 2005. *Catastrophe Modeling: A New Approach to Managing Risk.* New York: Springer.

Gurm, H., and D. Litaker. 2000. "Understanding the Influences on Informed Consent; Is 99% Safe Same as a Risk of 1 in 100?" *Academic Medicine* 75(8):840-842.

Haimes, Y. 1998. *Risk Modeling, Assessment and Management.* New York: Wiley.

Heal, G., and H. Kunreuther. 2006. "You Can Only Die Once: Interdependent Security in an Uncertain World." In H.W. Richardson, P. Gordon, and J.E. Moore II (eds.), *The Economic Impacts of Terrorist Attacks.* Cheltenham, U.K.: Edward Elgar.

Huber, O., and R. Wider. 1997. "Active Information Search and Complete Information Presentation in Naturalistic Risky Decision Tasks." *Acta Psychologica* 95(1):15-29.

Kasperson, R., N. Jhaveri, and J. Kasperson. 2001. "Stigma and the Social Amplification of Risk: Toward a Framework of Analysis." Chapter 2 in J. Flynn, P. Slovic, and H. Kunreuther (eds.). *Risk Media and Stigma.* London, U.K.: Earthscan.

Keeney, R.L. 1992. *Value-Focused Thinking.* Cambridge, Mass.: Harvard University Press.

Kunreuther, H. 2002. "Risk Analysis and Risk Management in an Uncertain World." *Risk Analysis* 22(4):655-664.

Lapan, H., and T. Sandler. 1988. "To Bargain or Not to Bargain: That is the Question." *American Economic Review* 78(2)16-20.

Lee, D. 1988. "Free Riding and Paid Riding in the Fight Against Terrorism." *American Economic Review* 78(2):22-26.

Loewenstein, G.F., E.U. Weber, C.K. Hsee, and N. Welch. 2001. "Risk as Feelings." *Psychological Bulletin* 127:267-286.

MacGregor D.G., P. Slovic, and T. Malmfors. 1999. How Exposed Is Exposed Enough? Lay Inferences About Chemical Exposure. *Risk Analysis* 19 (No. 4, August):649-659.

Michel-Kerjan, E. 2003. "Large-Scale Terrorism: Risk Sharing and Public Policy." *Revue d'Economie Politique* 113(5):625-648.

NRC (National Research Council). 1996. *Understanding Risk: Informing Decisions in a Democratic Society.* Washington D.C.: National Academy Press.

NRC. 2006. *Terrorism and the Chemical Infrastructure: Protecting People and Reducing Vulnerabilities.* Washington D.C.: The National Academies Press.

Parnell, G., R. Dillon, and T. Bresnick. 2005. "Integrating Risk Management with Homeland Security and Anti-Terrorism Resource Allocation Decision-Making." Pp. 431-461 in David Kamien (ed.), *The McGraw-Hill Homeland Security Handbook.* New York: McGraw-Hill.

Pillar, P. 2001. *Terrorism and U.S. Foreign Policy.* Washington D.C.: Brookings Institution Press.

Shipley, B. 2000. *Cause and Correlation in Biology: A User's Guide to Path Analysis, Structural Equations and Causal Inference.* Cambridge, U.K.: Cambridge University Press.

Slovic, P. 2000. *The Perception of Risk.* London, U.K.: Earthscan.

Tversky, A., and D. Kahneman. 1973. "Availability: A Heuristic for Judging Frequency and Probability." *Cognitive Psychology* 5(2):207-232.

von Winterfeldt, D. 1987. "Value Tree Analysis: An Introduction and Application to Offshore Drilling." Chapter 11 in P. Kleindorfer and H. Kunreuther (eds.). *Insuring and Regulating Hazardous Materials: From Seveso to Bhopal.* New York: Springer-Verlag.

The White House. 2004. Homeland Security Presidential Directive 10 [HSPD-10]: *Biodefense for the 21st Century.* Available at www.fas.org/irp/offdocs/nspd/hspd-10.html. Accessed January 16, 2008.

3

Description and Analysis of the Department of Homeland Security's Biological Threat Risk Assessment of 2006

> [T]he United States requires a continuous, formal process for conducting routine capabilities assessments to guide prioritization of our on-going investments in biodefense-related research, development, planning, and preparedness.
>
> —Homeland Security Presidential Directive 10: *Biodefense for the 21st Century*, 2004

The Department of Homeland Security's (DHS's) system for Biological Threat Risk Assessment (BTRA) is a computer-based tool that has been applied by DHS to assess the risk associated with the intentional release of each of the 28 biological agents listed in Figure 3.1. The methodology, an instance of probabilistic risk assessment (PRA), is described in *Bioterrorism Risk Assessment,* a report from the DHS Biological Threat Characterization Center of the National Biodefense Analysis and Countermeasures Center (DHS, 2006).

DHS credits seminal work on nuclear reactor safety as the basis for its risk assessment, citing "NUREG-1150" (case studies of probabilistic risk assessment) (U.S. Nuclear Regulatory Commission, 1991) and "NUREG-1489" (a tutorial on probabilistic risk assessment) (U.S. Nuclear Regulatory Commission, 1994) as basic references. The committee also found valuable an earlier foundation work, "NUREG 75/014" (U.S. Nuclear Regulatory Commission, 1975), widely known as the Rasmussen Report, which establishes the theoretical and policy foundations on which the 1991 and

CDC Category A Agents: (9 agents)	CDC Category B Agents: (15 agents)	CDC Category C Agents: (3 agents)	Genetically Engineered Agents: (1 agent)
• *Bacillus anthracis* • *Clostridium botulinum* toxin • Ebola virus (a VHF) • *Francisella tularensis* • Junin virus (a VHF) • Lassa virus (a VHF) • Marburg virus (a VHF) • Variola major • *Yersinia pestis*	• *Brucella suis* • *Burkholderia mallei* • *Burkholderia pseudomallei* • *Chlamydia psittaci* • *Clostridium perfringens* epsilon toxin • *Coxiella burnetii* • *Cryptosporidium parvum* • Eastern equine encephalitis virus • *Escherichia coli* O157:H7 • *Rickettsia prowazekii* • Ricin • *Salmonella typhi* • Shigella toxin • Staphylococcal enterotoxin B • *Vibrio cholerae*	• Bovine Spongiform Encephalopathy • Nipah virus • Rift Valley Fever virus	• MDR *Bacillus anthracis*

FIGURE 3.1 Biological threat agents as categorized by the Centers for Disease Control and Prevention (CDC). High-priority, Category A agents include organisms that pose a risk to national security because they can be easily disseminated or transmitted from person to person, they result in high mortality rates and have the potential for major public health impacts, they might cause social disruption, and they require special action for public health preparedness. Category B, the second-highest priority, includes agents that are moderately easy to disseminate, that result in moderate morbidity rates and low mortality rates, and that require specific enhancements of CDC's diagnostic capacity and enhanced disease surveillance. Category C agents include emerging pathogens that could be engineered for mass dissemination in the future because of availability, ease of production and dissemination, and potential for high morbidity and mortality rates and for major health impact. A later CDC-categorized list (CDC, 2007) features the same categories, but with agent entries revised. SOURCE: Available at www.bt.cdc.gov/Agent/Agentlist.asp.

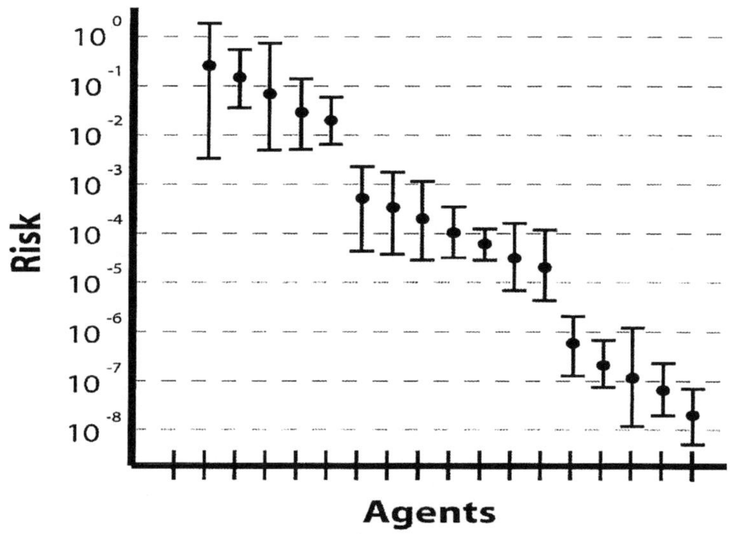

FIGURE 3.2 Ranking the risk of bioagents—the principal product of the Biological Threat Risk Assessment (BTRA) of 2006. In this figure, biological agents versus normalized risk, a sample display is based on fictitious data that represents only the general appearance of a key BTRA result. One of the vertical bars in this sample display represents anthrax; the dot shows the mean expected fatalities, and the horizontal bars show the 5th and 95th percentiles. However, as is done in the analyses included in DHS (2006), the vertical scale has been normalized so that the sum of the mean risks over all agents is 1. The committee does not know the normalization constant applied by BTRA and so cannot recover the actual expected risks.

1994 U.S. Nuclear Regulatory Commission reports and later applications depend.

The principal product of the BTRA of 2006 was a ranking of the risk posed by bioagent use based on calculated probabilities of expected fatalities. DHS chose to assess threat by ranking bioagents because government stakeholders had advised DHS that they "expected the primary assessments to be in the form of risk-prioritized groups of biological threat agents" (DHS, 2006, Ch. 1). Although a terrorist's choice of agent is just one step in a sequence of events leading to a potential attack, for practical purposes the BTRA of 2006 evaluates each agent separately. A probability is computed for each scenario involving that agent. Risk is then calculated as the product of these probabilities and the associated consequences. The overall risk associated with each agent is the integrated risk distribution over all possible scenarios involving that agent.

The product of the analysis by the BTRA of 2006 is displayed in a figure (such as Figure 3.2) that shows, for each agent, a normalization (whose normalization constant is not defined) of three estimated parameters of the distribution of consequences of agent attack in terms of expected fatalities:[1]

- The 5th percentile,
- The expected value (or mean), and
- The 95th percentile.

[1]The analyses presented in DHS (2006) are based entirely on estimated fatalities. However, DHS has conducted assessments based on illnesses and direct economic consequences as well.

For each agent, the estimate of the 5th percentile and of the 95th percentile of expected fatalities is displayed as a tick mark on a vertical line on a logarithmic ordinate scale of (normalized) consequences. The mean of expected fatalities is displayed as a dot. A typical display shows 28 parallel vertical lines, one for each agent. The specific numbers and rankings of agents by risk are functions of the assumptions underlying each of the many steps in the model's execution.

Before results are presented in DHS (2006), a normalizing constant is computed by multiplying, for each agent, the conditional expected consequence of the agent's use by the probability of its use, and then summing over all the agents. All statistics are divided by this constant to force the normalized means to sum to 1. This critical normalization constant is not displayed in the DHS (2006) report, so no *absolute* (versus relative) consequence can be recovered from the analysis presented there. Therefore, the normalization method cannot be verified by the committee. The normalization step is a curious one, in that it damages the results irreparably for purposes of decision making about, for instance, risk management. The committee conjectures that the normalization may reflect a well-intentioned but nonetheless an unfortunate effort to mitigate the stark nature of the estimated risks reported.

DHS (2006) also contains some qualitative analysis distinguishing between most-, less-, and least-"worrisome" bioagents. As for the quantitative analyses, consequences include only immediate numbers of expected fatalities. Future assessments have been promised with estimated casualties

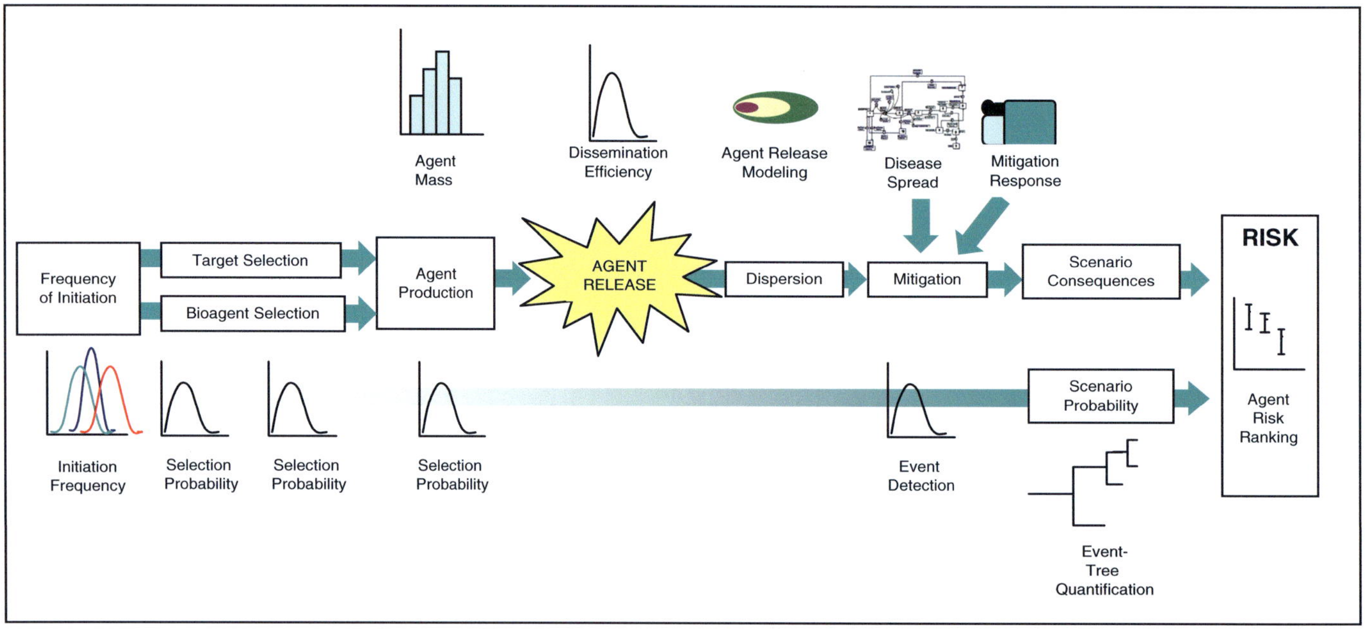

FIGURE 3.3 Biological Threat Risk Assessment (BTRA) event-tree risk assessment (left-to-right sequence) and consequence evaluation (at the right) are loosely coupled components. SOURCE: Tracy Hale, Battelle Memorial Institute, "2008 DHS Bioterrorism Risk Assessment: Planned Improvements," presented to this committee on February 10, 2007, Washington, D.C.

and indirect economic consequences. The committee does not know whether such estimates will also be normalized, but it hopes not.

DETAILS OF THE MODEL USED TO PRODUCE THE DEPARTMENT OF HOMELAND SECURITY'S BTRA OF 2006

The process that produced the estimates in the BTRA of 2006 consists of two loosely coupled analyses: (1) a PRA event-tree evaluation and (2) a consequence analysis (Figure 3.3). DHS has conducted "Material Threat Assessments" for single bioagents. "These are plausible, high consequence scenarios used to estimate the potential number of exposed individuals, their exposure levels, contaminated areas, and other collateral effects."[2] Presumably, the results of these assessments were used to inform the BTRA of 2006, but the committee was not briefed on them. DHS (2006) does not contain mathematical definitions of all of the parameters and variables used in the BTRA and does not present a complete mathematical model. (A complete mathematical model would show how each input is used to produce each output.) In response to the committee's request for that information, DHS has developed a lexicon and a mathematical model. Informed by discussions with DHS analysts, the committee's understanding of the details of the BTRA of 2006 is presented in this chapter using its own technical lexicon (Appendix A), which cross-references the terms used by the committee and those used in the DHS lexicon, when relevant. Readers interested more in the policy implications and potential uses of BTRA than in the technical details might want to skim the text of this chapter and read the four recommendations interspersed in the text below.

The BTRA of 2006 Uses a Probabilistic Risk Assessment Event Tree

A PRA event tree represents a sequence of random variables, called events, or nodes. Each random-event branching node is followed by the possible random-variable realizations, called outcomes, or arcs, with each arc leading from the branching, predecessor node, to the next, successor-event node (and it can be said without ambiguity that the predecessor event selects this outcome, or, equivalently, selects the successor event). With the exception of the first event, or root node, each event is connected by exactly one outcome of a preceding event. A node with no successor event is called a final event, or leaf. From each event, it is possible to trace a unique path back through alternating predecessor outcomes and events to the root event. The path from the root to a particular leaf is called a scenario. Each successive random event in a scenario path has a probability depending on all preceding outcomes in the path, and the probability of this scenario is the joint probability of the intersection of the outcomes on the path and is the product of these outcome probabilities. A natural way to construct an event tree is to

[2]John Vitko, Jr., Director, Chemical and Biological Division, DHS, Science and Technology Directorate, briefing to the BioShield Stakeholders Workshop, December 26, 2006.

place events in the chronological order in which they occur, if this order is known (e.g., Paté-Cornell, 1984).

This committee's concise mathematical definition of the BTRA event tree and associated computations are given in Appendix B.

Figure 3.3 shows some of the events in the BTRA tree. The "Frequency of Initiation" box at the extreme left consists of only one event—the beginning of a terrorist attack, which includes the terrorist's choice of frequency of attack, a random variable with four possible outcomes. Each frequency selected leads to a new event in the "Target Selection" box, as shown in Figure 3.3, and each of these four events is a random variable with eight possible outcomes, leading to a total of 32 events in the "Bioagent Selection" box. Each of these events is a random variable with 28 possible outcomes, depending on which of the 28 agents is used. Although not shown in Figure 3.3, there is a sequence of 17 such boxes in the BTRA event tree, enumerated and named in Figure 3.4, with each box corresponding to a different stage in the chronology of a terrorist attack. A complete listing of all the possible outcomes for these random variables is given in the BTRA documentation but not in this report of the committee. In the remainder of this chapter, the committee uses the term "stage" to mean all of the possible events at each step. As can be inferred from the names given to the stages (see Figure 3.4), each corresponds either to a terrorist decision (e.g., Bioagent Selection), or to a U.S. decision (e.g., Mitigation). It is a fundamental property of the BTRA of 2006 that every event, whether representing a terrorist decision or a U.S. decision, has a probability of occurrence associated with it.

As indicated above, Figure 3.4 displays the succession of 17 stages of the BTRA event tree. The BTRA represents *epistemic uncertainty* (uncertainty due to incomplete knowledge) by using a distribution of event probabilities from which a particular probability is sampled; that is, adopting the convention that from a node, each branching outcome "selects" a successor event, each such event leading to an outcome has a probability distribution over its probability of selection. For events in all but the first stage, each event leading to an outcome is chosen with a probability drawn from a distribution of probabilities for that outcome. The selection of outcomes from the only event in the first stage, "Frequency of Initiation by Terrorist Group," is the rate at which terrorists are anticipated to make attempts during a time horizon over which this rate applies; each such rate and time horizon has an associated probability.[3]

An 18th stage has been added to Figure 3.4 by the committee to represent the "Consequences" random variable. If the probability of an outcome depends on outcomes from an event in a preceding stage, the prior stage number is shown in column 3. The number of possible outcomes for each event in a stage is shown in column 4. The maximum cumulative number of paths into each stage is shown in column 5. Because outcome probabilities are conditional upon some preceding outcomes, column 6 shows the maximum number of such dependencies—this helps convey the complexity and sheer number of probabilities that must be reckoned for the BTRA.

In practice in the BTRA, the event tree is not actually evaluated as shown in Figure 3.4; each of the 28 agents (outcomes of events in Stage 3) is analyzed in isolation, yielding 28 sets of, in theory, as many as 350 million paths based on as few as 5,448 distinct probabilities for each agent. Although the maximum number of possible scenario paths is large (i.e., exponential in problem size), agent-by-agent, the event tree has many paths terminated early with no attack (e.g., by failure to manufacture an agent, by successful interdiction, and so on), while others continue to completion. Among the 28 event trees, each corresponding to the selection of a different agent, DHS (2006) reports one agent with only 1,184 scenarios, and another, the largest agent tree, with 192,928 scenarios.

The individual agent results are merged a posteriori into a distribution using probabilities for the selection of each agent and target. With the exception of this separation of event trees by agent, BTRA treats each of these successive events in ascending order of the stage in which it occurs.

For example, Figure 3.5 shows the outcomes for each event in Stage 2. After the frequency of attack has been chosen, the terrorist can choose among eight types of target to pursue. The BTRA represents the selection of each such outcome as an arc chosen randomly, with a selection probability that may depend on outcomes of events in prior stages. In this example, the outcome probabilities from events in Stage 2 may depend on the outcomes chosen for prior events in Stage 1.

The BTRA analyzes each of the 28 agents as follows:

1. The selection probability of the agent under study is set to 1 for each event in Stage 3. All other probabilities for events in Stage 3 are set to 0. (Stage 3 consists of agent-selection events; there are 32 events that result in agent-selection outcomes.) It is important to note that no attack using multiple agents is considered.
2. The tree for this agent is Monte Carlo generated, with outcome probability distributions conditioned upon outcomes from events in Stages 1 and 2 as well as on the knowledge of which agent is being modeled. BTRA represents epistemic uncertainty by using a distribution of outcome probabilities from which a particular probability is sampled. These epistemic probability distributions over outcome probabilities are elicited from subject-matter experts for each individual possible outcome, although there are thousands of such conditional outcomes.

[3]Given that the number of opportunities for such attempts is huge and the probability that any particular opportunity will be pursued is tiny, this is a Poisson rate.

Stage No.	Event Type	Depends on Stage No.	Number of Possible Outcomes	Maximum Cumulative Number of Paths into Stage	Maximum Number of Dependencies	Phase
1	Frequency of Initiation by Terrorist Group		4	4	4	Agent/Target/Dissemination Selection
2	Target Selection	1	8	32	32	
3	Bioagent Selection	2	28	896	224	
4	Mode of Dissemination (also determines wet or dry dispersal form)	1, 2, 3	9	8,064	8,064	
5	Mode of Agent Acquisition	3	4	32,256	112	Acquisition
(6)	Interdiction during Acquisition	1, 3, 5	2	64,512	896	
7	Location of Production and Processing	1	2	129,024	8	Production and Processing
8	Mode of Agent Production	1, 3	3	387,072	336	
9	Preprocessing and Concentration	1, 2, 3, 4, 8	3	1,161,216	72,576	
10	Drying and Processing	1, 2, 3, 4	3	3,483,648	24,192	
11	Additives	1, 2, 3, 4	2	6,967,296	16,128	
(12)	Interdiction During Production and Processing		2	13,934,592	2	
13	Mode of Transport and Storage	1, 2, 3, 4	3	41,803,776	24,192	Transport and Storage
(14)	Interdiction During Transport and Storage	7	2	83,607,552	4	
(15)	Interdiction During Attack		2	167,215,104	2	Attack
16	Potential for Multiple Attacks	1	2	334,430,208	8	
(17)	Event Detection	2, 3, 4	3	1,003,290,624	6,048	Response
18	Consequences	tbd	10	10,032,906,240	tbd	Final Outcome

FIGURE 3.4 Successive stages in the Biological Threat Risk Assessment (BTRA) event tree. A BTRA event tree consists of 17 stages classified into six successive phases. The committee has emphasized Stages 6, 12, 14, 15, and 17 by inserting parentheses around these stage numbers in the left-hand column, to distinguish interdiction opportunities. Outcomes of events in all other stages are chosen by the bioterrorist. The committee added the columns labeled "Number of Possible Outcomes," "Maximum Cumulative Number of Paths into Stage," and "Maximum Number of Dependencies," as well as an 18th stage representing "Consequences." NOTE: tbd, to be determined. SOURCE: Adapted from DHS (2006, Table 5.1).

Stage No.	Event Type	Possible Outcomes	Depends on Stage No.
2	Target Selection	2.1 Large Open Building	1
		2.2 Small Enclosure	
		2.3 Large "Divided" Building	
		2.4 Large Outdoor Spaces	
		2.5 Water Pathway	
		2.6 Food Pathway	
		2.7 Human Vectors	
		2.8 Contact (letters)	

FIGURE 3.5 Each event offers the terrorist one choice of a number of alternate outcomes. Here, Stage 2, "Target Selection," is amplified into eight outcomes. The Biological Threat Risk Assessment represents the choice of each outcome with a probability and refers to this as a "split fraction" (i.e., conditional arc probability). The number at the right shows that the probability distribution on outcomes from events in Stage 2 is dependent on outcomes from events in Stage 1, "Frequency of Initiation by Terrorist Group." SOURCE: Adapted from DHS (2006, Table 5.2).

3. A set of outcome probabilities is generated, and the resulting probabilistic risk assessment event tree is solved. That is, each leaf (terminal event) with nonzero probability is associated with a consequence distribution, from which the leaf-probability-weighted consequence distributions are sampled to produce a sample unconditional consequence distribution. The BTRA does this 500 times, thus generating a random sample of 500 PRA trees and associated consequence distributions. For each of these trees, the resulting 5th and 95th percentiles and the average of the consequences are computed. This sampling of multiple realizations from the same starting conditions represents *aleatory uncertainty*—the influence of pure randomness.
4. The outcome of each random-sample scenario is captured by the distribution of *expected* consequences; the expectation is over purely aleatory randomness.
5. The 28 agent statistics are merged, after the fact, using the agent-selection probabilities.

The committee's hypothetical scenario, introduced in Chapter 1, may be approximately described by a number of possible sequences of outcomes in the BTRA event tree. The type of terrorist group here would not likely be a deranged individual or even a small cell, because the volume of anthrax hypothesized for this large-scale, outdoor aerosol attack exceeds that of the attacks following 9/11 by several orders of magnitude, and thus the terrorists are evidently well funded, perhaps even state-sponsored. Target selection (Figure 3.5) would be a "Large Outdoor Space." That this space is hypothetically filled with commuters conditions the consequences, but it is not clear where these commuters would appear in the BTRA; they are evidently rolled up along with a host of other considerations for subject-matter experts to consider when rendering opinions about consequences. Event by event, outcomes that support this scenario can be identified, although many nuances (e.g., steps to concentrate, process, and introduce additives to "weaponize" the anthrax spores for better dispersal) may be hard to unambiguously identify (i.e., the attackers have either weaponized a lot of anthrax, or they have produced an even larger quantity of crude anthrax to use). Regardless, the base mission of the BTRA is to automatically generate hosts of scenarios, including ones that resemble the committee's hypothetical scenario, and rank them in terms of expected risk (i.e., fatalities).

Three short papers (DHS, 2007a,b,c) presented to the committee give details on and contain versions of Figure 3.6. In this tree, the starting event is at the extreme left, followed by two stages of events representing the terrorist choice of agent and then choice of target. A complete scenario in this reduced example is characterized by a left-to-right scenario path from starting event to final event and is documented by the successive outcomes, or arcs, in this scenario path. For instance, a path with arcs labeled "P_{A1}, $1-P_{T1}$" leads to scenario s_2 with consequence distribution $c(x \mid s_2)$, where x represents fatalities. The notation "P_{A1}" represents, at once, the selection of Agent 1 and its probability of selection. Although not shown in Figure 3.6, each successive probability could depend on everything that precedes it in its scenario path. So, in example scenario path "P_{A1}, $1-P_{T1}$," the probability P_{T1} can depend on the prior choice of event P_{A1}.

As noted above, a fundamental property of the event trees used in the BTRA is that every decision by a bioattacker (e.g., choice of an agent) or by a defender (e.g., choice of a countermeasure) is considered to be an uncertain event—hence associated with an outcome selection probability. In

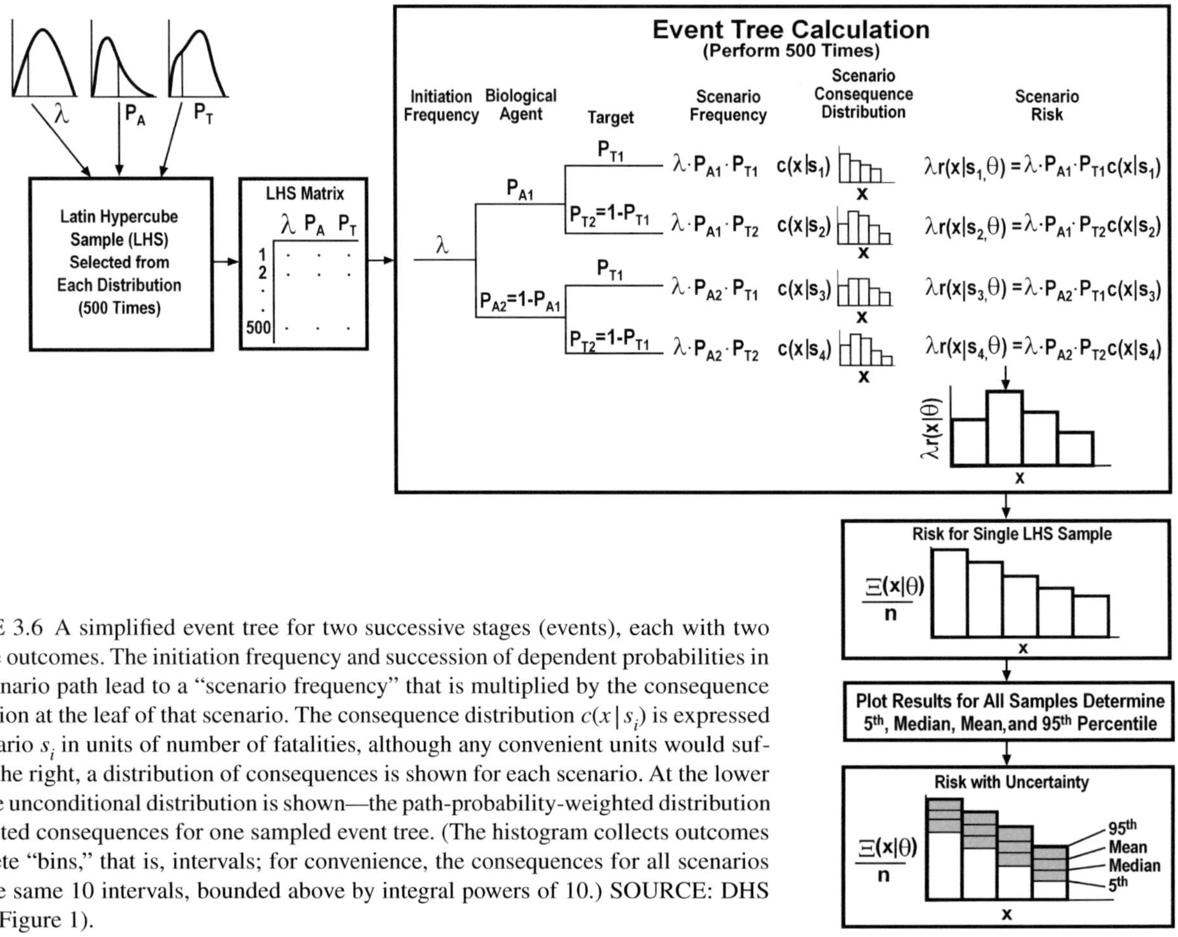

FIGURE 3.6 A simplified event tree for two successive stages (events), each with two alternate outcomes. The initiation frequency and succession of dependent probabilities in each scenario path lead to a "scenario frequency" that is multiplied by the consequence distribution at the leaf of that scenario. The consequence distribution $c(x \mid s_i)$ is expressed for scenario s_i in units of number of fatalities, although any convenient units would suffice. At the right, a distribution of consequences is shown for each scenario. At the lower right, the unconditional distribution is shown—the path-probability-weighted distribution of expected consequences for one sampled event tree. (The histogram collects outcomes in discrete "bins," that is, intervals; for convenience, the consequences for all scenarios share the same 10 intervals, bounded above by integral powers of 10.) SOURCE: DHS (2007c, Figure 1).

fact, the BTRA uses pure probability trees, and no decision tree at all. A statement in a presentation to the committee: "An event tree (decision tree) is a visual tool . . ."[4] indicates confusion on this point. The distinction between event and decision trees is fundamental, not semantic. In event trees, all outcomes are modeled as random events determined by some probability distribution; decision trees allow the possibility that outcomes are chosen by the defender or attacker to achieve some objective. Decision trees as tools for modeling terrorist threats are discussed in Chapter 7.

In step 3 above, for each outcome from each event, the probability of selection has been elicited as the consensus of a group of subject-matter experts in the form of an expected probability, and reportedly some additional guidance (such as the 5th and 95th percentile of this outcome probability) that has been transformed by some unspecified means into a variance for each probability. Each of these outcome selection probability solicitations is converted into a marginal probability density of probabilities for selecting the particular outcome. Documentation indicates that most subject-matter experts for the BTRA of 2006 were experts from Battelle Memorial Institute, Columbus, Ohio, but that subsequent work will draw from a much wider pool of experience.

Some observations by the committee about the details in these steps follow. In step 2 above, the Monte Carlo simulation generates probabilities for each event *one outcome (arc) at a time in some fixed sequence of outcomes.* For each successive outcome, a marginal probability distribution over the probability for selecting this outcome is used. The probability distribution for each successive outcome is *conditioned* on the probabilities already realized for this event. Because the outcome probabilities must sum to 1, the marginal distributions for each should be constrained to have their expectations sum to 1. Although the original marginal distributions are, for instance, beta densities, the successive conditioning by sampled outcomes means that outcomes are really sampled from some multivariate density for which the marginals are not beta, and in fact are not

[4] Richard S. Denning, Battelle Memorial Institute, "DHS 2006 Bioterrorism Risk Assessment Methodology," presentation to the committee, August 28, 2006, Slide 8.

characterized at all in closed form, and not mentioned at all in DHS (2006).

The DHS procedure selects the last outcome probability so that the sum of outcome probabilities emanating from this event is 1 (i.e., the last marginal probability distribution is not used at all). However, the outcome probabilities should have a *joint* distribution that captures their dependencies (the most important being that they sum to 1). Even if the present method were not technically superfluous (as is shown below), subject-matter experts typically cannot assess such high-dimensional distributions (Moskowitz and Sarin, 1983).

In step 3, the 500 sets of outcome probabilities for each agent event tree are obtained using a Latin Hypercube Sampling design (Stein, 1987), a sampling technique applied in earlier years to probabilistic risk analysis of nuclear safety. However, the committee notes that this sampling design produces unbiased estimates of the mean and quantiles *with asymptotic sample size*. Further, see Stein (1987, p. 144, Equation (3) and Section 5) and McKay et al. (1979, Section 8.3). Moreover, the variance may be decreased *or increased* by this design, depending on the covariance structure of the distributions sampled. Note that the proofs of unbiasedness for quantiles are for independent random variables. There is no evidence that the efficacy of the particular BTRA sample design has been established.

The BTRA of 2006 Does Not Use Event Trees for Consequence Analysis

Consequence models characterize the probability distribution of consequences for each scenario. The BTRA employs a mass-release model that assesses the production of each bioagent, beginning with time to grow and produce, preprocess and concentrate, dry, store and transport, and dispense. The net result is a biological agent dose that is input to a consequence model to assess casualties. One equation from the model is produced here to give a flavor of the computations.

$$MR = MT \times QF_1 \times QF_2 \times QF_3 \times QF_4 \times QF_5$$

where MR is bioagent mass release, MT is target mass, and QF_i are factors to explain production, processing, storage, and so on and are random variables conditioned on the scenario whose consequences are being evaluated.

The complete model computes, for an attack with a given agent on a given target, how much agent has been used, how efficiently it has been dispersed (and, for an infectious agent, how far it spreads in the target population), and the potential effects of mitigation efforts. For the BTRA of 2006, all of these factors were assigned values by eliciting opinions of subject-matter experts in the form of subjective discrete probability distributions of likely outcomes, and by some application of information on the spread of infectious agent, atmospheric dispersion, and so on.

The BTRA consequence analysis is qualitatively different from its event-tree analysis. Subject-matter expert opinions are developed much like case studies, and there is less clear dependence on specific events leading to each consequence. Thus, each consequence distribution should be viewed as being dependent on *every* event leading to its outcome. However, an examination of the underlying analysis in the DHS (2006) report suggests that there is really only a *single* consequence distribution for each scenario: one that depends not on the complete scenario but only on a subset of parameter values. (Indeed, "Consequence uncertainty was omitted due to the overwhelming processing requirements."[5]) A Monte Carlo simulation of 1,000 samples was used to estimate each consequence distribution in the BTRA of 2006. The committee has no details about how this was accomplished.

THE EVENT TREE CAN BE IMPROVED

The Approach to Determining the Probabilities of Terrorist Decisions Is Incomplete

The BTRA of 2006 uses probabilities to represent adversarial decisions. These are conditional probabilities, but the conditioning is retrospective, rather than prospective. Consider that if the consequence model for a bioagent is completely changed to reflect some new discovery about the efficacy of the bioagent, this would have no influence at all on the BTRA probabilities; neither the terrorist nor the United States would change probabilities in response.

When dealing with an intelligent, goal-oriented, and resourceful adversary, not with a force such as nature that randomly determines whether unwanted events occur, this committee believes that the use of probabilities to represent bioterrorism decisions must be tempered by a thorough understanding of how these probabilities have been assessed (whether by means of formal game-theoretical models, elicitation of subject-matter experts, of other means). For decision problems as complex as those motivating BTRA, the assessment of the probabilities that adversaries will choose courses of action should be the *outputs* of analysis, not required *input parameters*. The BTRA has reversed this preferred approach by requiring that subject-matter experts predict, a priori, how adversaries will behave. For this approach to make sense, the subject-matter experts must grasp nuances of alternatives and outcomes and render opinions founded on an analysis of the entire decision process, which would be very difficult for a process this complex. The committee saw no evidence that this level of analysis was used. Moreover, the static probabilities used are not appropriate when terrorists can observe and react dynamically to any earlier decisions made by the United States.

[5]Traci Hale, Battelle Memorial Institute, "2008 DHS Bioterrorism Risk Assessment: Planned Improvements," presented to this committee on February 10, 2007, Washington, D.C.

Recommendation: To assess the probabilities of terrorist decisions, DHS should use elicitation techniques and decision-oriented models that explicitly recognize terrorists as intelligent adversaries who observe U.S. defensive preparations and seek to maximize the achievement of their own objectives.

It should be noted that this recommendation does not require the prediction of terrorist actions, a difficult task at best. Its intent is to evaluate risk on the basis of hypothetical terrorist attacks against U.S. defenses that have been designed to thwart terrorist goals. Thus, its implementation will produce a conservative estimate of risk. In Chapter 7, the committee offers alternate modeling techniques to accomplish this more complex assessment.

The Mathematics Used by the BTRA in Modeling Multiple Attacks Has Errors

Given a successful attack, the PRA tree's Stage 16, Potential for Multiple Attacks[6] (see Figure 3.4) presents an opportunity for the terrorist to mount more such attacks. The probability for succeeding at each additional attack is given as λ' (implying that the attacks that are attempted first are no more likely to succeed than those postponed until the first attempts have failed), and the expected number of attacks before interdiction is given in the DHS (2006) report and presentation to the committee as

$$f_1(\lambda') = 1 + \frac{\lambda'}{(1+\lambda')^2}.$$

This expectation is *multiplied* by the consequence distribution for such attacks.

During a site visit to Battelle in Columbus, Ohio, in October 2006, the committee pointed out that this equation must be in error (e.g., if $\lambda' = 1$, the expected number of re-attacks should go to infinity, but $f_1(\lambda' = 1) = 1.25$).

Subsequent briefing materials. (Battelle Columbus Operation, 2007) featured a new expectation:

$$f_2(\lambda') = 1 + \frac{\lambda'}{(1-\lambda')^2}.$$

This expectation is also wrong. Given one successful attack, the total number of successful attacks before an interdiction with probability of success for each additional attack λ' is

$$f_3(\lambda') = 1 + E[n \mid \lambda'] =$$

$$1 + \sum_{n=0}^{\infty} n(\lambda')^n (1-\lambda') = 1 + \frac{\lambda'}{1-\lambda'} = \frac{1}{1-\lambda'}.$$

Figure 3.7 shows these expressions as a function of λ'. This has a significant influence on the expected consequences of multiple attacks. For $\lambda' = 0.9$, $f_1(0.9 = 1.25$, $f_2(0.9) = 91$, and the correct expectation $f_3(0.9) = 10$. For this numerical example, the two expectations respectively would underestimate and overestimate consequences by an order of magnitude.

If each repeated attack is independent, the distribution of total consequences across all attacks will presumably be additive. *The distribution of this sum is characterized by a statistical convolution, not by mere multiplication by the expected number of re-attacks.*

It is not very realistic to assume an infinite supply of potential attacks that all have equal probabilities of success. Judging from U.S. actions taken after the 9/11 terrorist attacks, the committee believes that all of the probabilities assessed in the event tree will change following any attack.

Thus, the implicit, homogeneous steady-state Poisson process underlying the rate used for "Frequency of Initiation by Terrorist Group" will almost surely be rendered invalid by any detected attack, whether successful or not, and whether interdicted or not. Subsequent to any such event, the BTRA analysis would be rendered inapplicable until a host of key parameters could be reestimated and the BTRA then repeated from scratch.

The BTRA multiple-attack feature is an embellishment that has been incorrectly implemented both mathematically and statistically, and even if correctly implemented would be based on a questionable underlying model.

The 2006 BTRA's Assessment of Outcome Probabilities Is Unnecessarily Complex

Each node in the PRA tree offers two or more outcomes leading to successor events, each selected with an epistemic probability density that is used to generate an aleatory outcome probability to be used to solve the event tree. Each of these outcome densities is typically a beta density function, formed somehow from averages elicited from subject-matter experts, whose means sum to 1. It is straightforward to show that, when given a distribution over outcome probabilities, *the means of these distributions suffice to completely capture the unconditional distributions over any consequence.* For example, suppose that

ξ_i = probability that outcome i will occur, $i = 1,2, \ldots, n$

$\vec{\xi} = [\xi_1, \xi_2, \ldots, \xi_n]$

$f_i(x)$ = probability distribution over outcome X, given that outcome i occurs.

If the values of ξ_i are known, then the *unconditional* distribution over consequences X is

$$f(x \mid \vec{\xi}) = \sum_{i=1}^{n} \xi_i f_i(x).$$

[6]"Multiple attacks" refer to attacks in sequence. "Simultaneous multiple attacks" are considered by the BTRA to be a single attack.

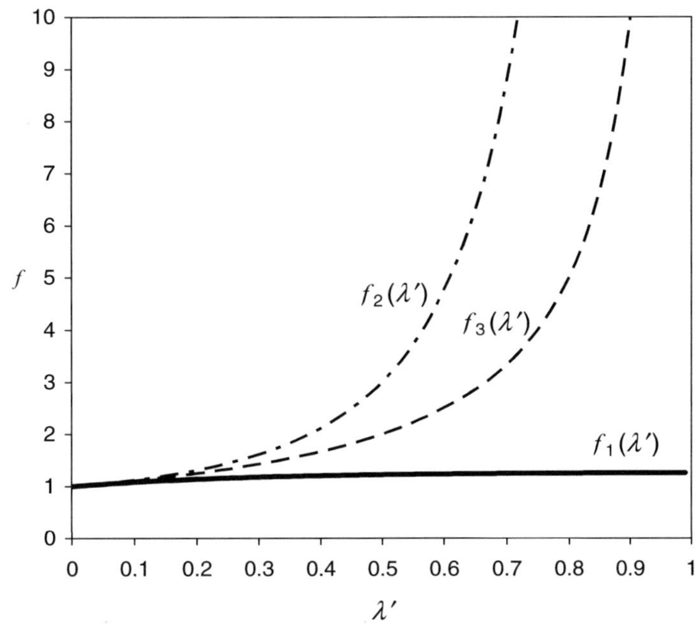

FIGURE 3.7 Expected number of attacks before interdiction, given that a first attack is successful and that continued attacks each evade interdiction with probability λ'. $f_3(\lambda')$ is the expected number of attacks before interdiction. f_1 is the BTRA expression, and f_2 is the expression offered with a complete numerical example (Battelle Columbus Operation, 2007). For $\lambda' = 0.9$, f_1 underestimates by an order of magnitude, and f_2 overestimates by an order of magnitude. This expectation is multiplied by the single-attack distribution of consequences, so these errors have major influence. The value $\lambda' = .9$ has been chosen by the committee for expository purposes. The committee does not represent that this value occurs in any scenario analyzed by DHS or in the example provided by Battelle. The interested reader my substitute any other value for λ' to assess its effect.

However, the epistemic approach considers ξ_i to be a random variable, and therefore $\vec{\xi}$ a random vector. Letting $\varphi(\vec{\xi})$ be the (joint) probability distribution over the elements ξ_i of the random vector $\vec{\xi}$, the unconditioned distribution over consequences becomes

$$f(x) = \int f(x \mid \vec{\xi})\varphi(\vec{\xi})d\vec{\xi} =$$

$$\int \sum_{i=1}^{n} \xi_i f_i(x)\varphi(\vec{\xi})d\vec{\xi} = \sum_{i=1}^{n} f_i(x)\int \xi_i \varphi(\vec{\xi})d\vec{\xi} = \sum_{i=1}^{n} f_i(x)E(\xi_i).$$

Therefore, even when using a general (and possibly highly dependent) joint distribution, all that is needed is the expectation $E(\xi_i)$, which is the mean of the epistemic distribution; the rest of the distribution is irrelevant to determining the unconditional distribution of consequences (and, in particular, its moments, percentiles, and so on).

Because of this, the consequence distribution can be calculated *without sampling* from the outcome probability distributions. Appendix C provides a self-contained, simplified example of this point. For an event tree the size of the one used in the BTRA of 2006, this represents a significant computational simplification and would also significantly simplify the BTRA exposition; both of these results are desirable. What is lost in the simplification is the family of risk curves—i.e., one curve for each possible outcome. However, no analysis in the BTRA of 2006 and no improvement in analysis recommended by the committee can make meaningful use of the information available in the family of risk curves, beyond that provided by their expectation. Further, given the planned improvements to the BTRA incorporating additional consequence measures and utility functions, the committee does not anticipate analyses that require the family of risk curves.

If the conditional consequence distributions are given in parametric form, or in numerical lookup tables, calculation of the risk distribution can be done exactly, without resorting to estimating these distributions from the outputs of Monte Carlo simulations. This computation is easy and fast, and the result is the distribution—not merely an estimate of its features.

For these reasons, the committee's finding is that the epistemic features of the BTRA probabilistic risk assessment are unnecessary and that they increase computation time and complicate exposition, analysis, and understanding of results.

Recommendation: The event-tree probability elicitation should be simplified by assessing probabilities instead of probability distributions for the outcomes of each event.

BTRA Results Should Not Be Normalized by an Unspecified Constant

The absence of a normalization constant in DHS documentation and presentation irretrievably obscures those BTRA results where normalization is employed, rendering those results essentially useless for further analyses, especially for risk management. As an illustrative example, suppose that the United States discovers how to make a reliable biological agent alarm the size and cost of a smoke detector and how to connect such detectors to local area networks; educates the U.S. populace to shelter in place on alarm; implements effective, immediate cordoning and quarantine procedures; and thus attains an estimated threefold reduction in expected consequences from terrorist use of all biological agents. *This improved capability of detection and response would not change a single normalized result presented in the BTRA of 2006.*

The committee wonders how senior leadership has interpreted a normalized fatality scale (with no units) in the DHS (2006) report and presentation materials: the committee does not know why this normalization was applied, and especially why its essential details are absent from all underlying documentation. The normalized results are classified, as would be the non-normalized results, and this one step—normalization—has made it impossible for anyone to reproduce any BTRA result or for anyone to use independent means to assess the accuracy of any BTRA result.

Most important, risk management deals with risk, not normalized risk. The BTRA needs to report risk, not normalized risk.

Recommendation: Normalization of BTRA risk assessment results obscures information that is essential for risk-informed decision making. BTRA results should not be normalized.

The BTRA Event Tree Can Be Simplified

The BTRA is a risk assessment. (The committee argues in Chapter 4 that mere risk assessment is inadequate, but for purposes of this chapter the committee adopts the purely probabilistic BTRA view with the objective of improving the exposition of the methodology used.) The committee thinks that the entire BTRA analysis can be envisioned, implemented, carried out, and documented more simply and clearly as a single, unified probabilistic risk analysis with a single PRA tree that includes conditional consequence distributions.

The fixed sequence of 17 stages (or 18, including the committee's additional stage) drives the BTRA. The analysis has been frustrated by the sheer size of the PRA tree for all biological threat agents, and as a practical matter, the BTRA separates the 28 agents and solves each PRA tree in isolation. But given that selection of agent is the third stage in the fixed hierarchy of the BTRA event sequence, and that this selection depends on both prior stages, this approach has complicated the analysis and exposition of results. Fixed adherence to the 17 sequential stages in the BTRA event tree leads to large PRA trees that have had to be separated by 28 individual agents. That the choice of agent is not a first-stage event, or even a second-stage one, but rather a third-stage event, causes some difficulty in recovering results after the fact.

Recommendation: Two significant simplifications should be made to the BTRA of 2006 event tree:

- **DHS should eliminate Stage 1, Frequency of Initiation [of an attack] by Terrorist Group, and Stage 16, Potential for Multiple Attacks; and**
- **DHS should seek opportunities to aggregate some stages of the tree to only those essential to calculate probabilities and consequences with realistic fidelity.**

The elimination of probability elicitation for terrorist decisions will greatly simplify the model. Additional simplifications are also possible. For instance, Stages 7 through 11 (successively: Location of Production and Processing, Mode of Agent Production, Preprocessing and Concentration, Drying and Processing, and Additives) appear to reflect a somewhat artificial taxonomy and permutation of decisions in a proliferation effort. Similarly Stages 14 and 15 (Interdiction During Transport and Storage, and Interdiction During Attack) might be aggregated. It should be noted, however, that the level of detail shown in Figure 3.4 may coordinate with the steps that the FBI and the National Counterterrorism Center consider in evaluating an attack, because in many cases these steps can be associated with specific technical capabilities and as a result can be tied to intelligence assessments of what capabilities and activities have occurred. The committee has received no briefings on this aspect of the evaluation of the bioterrorism threat.

ADDITIONAL OBSERVATIONS REGARDING THE DEPARTMENT OF HOMELAND SECURITY'S BTRA OF 2006

Reporting Results

According to DHS (2006), given the use of a particular agent, the probability of a typical scenario on the BTRA event tree may be on the order of 10^{-10}. A typical end-state consequence may be on the order of tens of thousands of fatalities or more. The product of this probability and consequence represents a particular scenario's contribution to the total expected consequence associated with the use of that agent. The sum of these represents that agent's "relative importance," given that it is selected for use. The result of multiplying these infinitesimal probabilities by large numbers of casualties yields risk, which is represented with the 5th percentile, mean, and 95th percentile of this risk, normalized by the total expected risk of all agents.

DHS need not focus exclusively on using the rank of relative (expected) risk (and presents only 3 statistics, i.e., 5th percentile, mean, and 95th percentile, per agent) as the final result of all this analysis. With a pure event tree, more information and insight can be obtained by a more thorough analysis.

It would be easy to illustrate on the same plot (similar to Figure 3.2, with a probability above or below each vertical bar representing an agent) that (for the BTRA of 2006), for example, the agent selection probability estimated by one subject-matter expert was greater than 40 percent, by another almost 30 percent, by a third greater than 10 percent, by a handful of others about 5 percent, with the rest much smaller. That is, one of the "most likely" three agents was selected with 80 percent probability. The committee believes that presenting the prior probability of agent selection—a key subject-matter expert opinion—on the same plot with the level of risk associated with the use of each particular agent would help determine whether these estimates by subject-matter experts are credible and would help interpret whether agent-selection probability is a significant factor leading to agent risk. This is extremely valuable information that is not easy for the reader to recover from analyses presented in DHS (2006) (and cannot be recovered at all from the Executive Summary of DHS [2006]).

Similarly, it would be easy to show the number of scenarios (i.e., successful attack paths) associated with each of the 28 agents. This and other simple gauges would lend insight into the robustness of each PRA tree with respect to either U.S. or terrorist decision represented in the tree.

"Prioritization" with a strict ranking by specific agent may not be the best way to present results. For instance, if one simple, cheap action can remediate the consequences of a number of infectious agents, none of which appears in the top tier of qualitatively identified "worst" ones, the rank-ordering would not reveal this. The BTRA of 2006 does not anticipate prescriptive covering of multiple-agent risks by a single action or set of actions. In subsequent chapters the committee recommends the pursuit of a resource-constrained optimization of DHS investments to maximize total risk mitigation, and suggests some examples in Chapter 7.

Tailored Risk Assessments

The BTRA of 2006 conducts a series of "tailored" risk assessments that address, in particular:

- *High-consequence (i.e., high-fatality) events.* Because these events are of keen concern to decision makers, consequence distributions are truncated below a threshold number of fatalities, and the conditioned risk rankings are presented.
- *Prioritization of agents for purposes of research.* This analysis seeks to identify particular discoveries that might have a large influence on risk.
- *Prioritization of agents for purposes of development of medical countermeasures.* This analysis seeks to identify improvements in medical countermeasures that could impact expected fatalities. "The metric for assessing the potential impact of countermeasure development research is based on two criteria: baseline fatality risk and current countermeasure efficacy. This prioritization does not consider the current state of research on each agent, i.e. how close current countermeasure research is to a countermeasure breakthrough on individual agents" (DHS, 2006, Ch. 3, p. 11).

Analysis of Sensitivity and Risk

The BTRA of 2006 offers an additional set of results that investigate (1) how much key assumptions contribute to the results of the risk analysis and (2) how much alternative risk mitigation strategies might reduce overall risk. Sets of runs systematically vary the epistemic outcome probabilities. Key assumptions are examined by varying parameters for agent selection and acquisition, production, and utilization; risk mitigations are examined by varying parameters related to interdiction and medical mitigation. In Chapter 4, the committee discusses the importance of sensitivity analysis and the difficulty of accomplishing sensitivity analysis with BTRA.

Critical Knowledge Gaps and Biodefense Vulnerabilities

Critical knowledge gaps provide the greatest opportunities for the reduction of uncertainty in risk analysis, while critical biodefense vulnerabilities provide the greatest areas for the reduction of risk. The BTRA of 2006 identifies three areas of critical knowledge gaps: (1) intelligence and terrorist organization preferences, (2) event detection and response, and (3) biological threat agent properties. The possibility of using these threat agent properties to aggregate current and potential biological agents is discussed in Chapter 5.

Two areas of critical biodefense vulnerabilities are examined: (1) threat-related vulnerabilities and (2) consequence-and/or mitigation-related vulnerabilities (Figure 3.8).

Planned Improvement for the BTRA of 2008

Homeland Security Presidential Directive 18: *Medical Countermeasures Against Weapons of Mass Destruction* (The White House, 2007) states:

> The Secretary of Homeland Security shall develop a strategic, integrated all-CBRN risk assessment that integrates the findings of the intelligence and law enforcement communities with input from the scientific, medical, and public health communities. Not later than June 1, 2008, the Secretary of

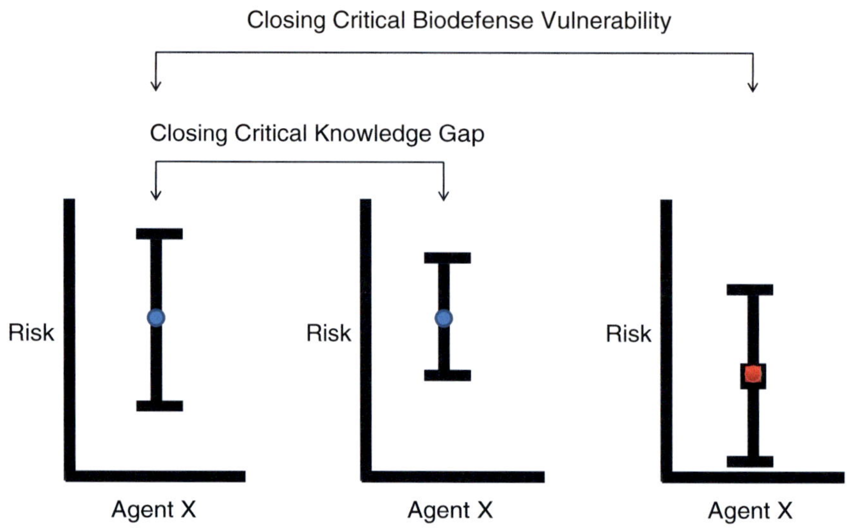

FIGURE 3.8 Closing a critical biodefense vulnerability reduces the overall risk but may not affect the uncertainties associated with that risk. Closing a critical knowledge gap does not reduce risk, but does lower the uncertainty associated with the risk. SOURCE: Adapted from DHS (2006, Figure 4.1).

> Homeland Security shall submit a report to the President through the Assistant to the President for Homeland Security and Counterterrorism, which shall summarize the key findings of this assessment, and shall update those findings when appropriate, but not less frequently than every 2 years.

With this guidance, the following BTRA activity was undertaken in 2007 to support the future DHS report on the BTRA of 2008. The committee offers a few comments, not anticipated in its charge, shown in italics.

- The consequence models will employ epistemic sampling, and there will be more than 10 consequence bins in the discrete consequence distributions.
- A library of consequence models will include a Leontief model of indirect economic consequences, a water contamination model, agricultural disease models, a differential equation model of the spread of infection and the effects of medical countermeasures, atmospheric dispersion forecasts, air circulation models within buildings, and others. The specific means by which outputs from these models will be converted into consequence distributions has not been presented to the committee. *Chapter 4 of the present report cautions against including excessive detail in these models where there are insufficient supporting data.*
- DHS plans to develop its own model of food supply contamination in cooperation with various other agencies and BTSafety, LLC,[7] and anticipates cooperating with the Environmental Protection Agency (EPA) to use EPA's existing models of waterborne contamination (EPA, 2007).
- DHS is developing a detailed susceptible, exposed, infected, and recovered (SEIR) model for the spread of infectious agents, using STELLA,[8] to simulate disease transmission and medical mitigation measures through the solution of systems of differential equations. *In Chapter 6, the committee cautions that there may be insufficient scientific knowledge to verify or validate these models.*
- In addition to indoor aerosol dispersion models, DHS is particularly interested in modeling an agent release and spread in a subway system.
- DHS plans to cooperate with the Lawrence Livermore National Laboratory (2006) and the National Center for Foreign Animal and Zoonotic Disease (FAZD) at Texas A&M University.[9]
- BTRA plans to incorporate more agents, including anti-agricultural, engineered, and emerging agents.

Although the committee agrees that some additional human-threatening agents and agricultural agents may warrant attention, the committee recommends *less* detail in future BTRA analyses, rather than more. Chapter 5 suggests aggregate categorization of agents. Such simplification would not materially damage model credibility or fidelity, given the enormous volume of assumptions and estimates required to instantiate any given event tree. Simplification

[7]For further information, see www.btsafety.com/software.htm. Accessed February 23, 2007.

[8]For further information, see www.iseesystems.com. Accessed February 23, 2007.

[9]For further information, see fazd.tamu.edu. Accessed February 23, 2007.

would yield more insights, accessible results, faster computation, and thus better responsiveness to requests for information, as addressed in Chapter 4. Insightful analysis explaining the "why" of results is much more important than additional detail cluttering the "what" of results.

REFERENCES

Battelle Columbus Operation. 2007. "Detailed Single Scenario Analysis." Prepared for National Biodefense Analysis and Countermeasures Center. Columbus, Ohio, March 27.

CDC (Centers for Disease Control and Prevention). 2007. "Emergency Preparedness and Response—Bioterrorism Agents/Diseases." Available at www.bt.cdc.gov/agent/agentlist-category.asp. Accessed April 13, 2007.

DHS (Department of Homeland Security). 2006. *Bioterrorism Risk Assessment.* Biological Threat Characterization Center of the National Biodefense Analysis and Countermeasures Center. Fort Detrick, Md.

DHS. 2007a. "A Lexicon of Risk Terminology and Methodological Description of the DHS Bioterrorism Risk Assessment." Draft document dated February 2, 2007. Washington, D.C.

DHS. 2007b. "Example of Risk Assessment Calculations." Document provided to the Committee on Methodological Improvements to the Department of Homeland Security's Biological Agent Risk Analysis. February 26, 2007. Washington D.C.

DHS. 2007c. "A Lexicon of Risk Terminology and Methodological Description of the DHS Bioterrorism Risk Assessment." Written communication to the Committee on Methodological Improvements to the Department of Homeland Security's Biological Agent Risk Analysis. April 14, 2007.

EPA (Environmental Protection Agency). 2007. *A Water Security Handbook: Planning for and Responding to Drinking Water Contamination Threats and Incidents.* Available at www.watersc.org/pdf/water_security_handbook_rptb.pdf. Accessed February 23, 2007.

Lawrence Livermore National Laboratory. 2006. "Protecting Our Nation's Livestock." *Science and Technology Review.* Available at www.llnl.gov/str/May06/Lenhoff.htm. Accessed February 23, 2007.

McKay, M., R. Beckman, and W. Conover. 1979. "A Comparison of Three Methods for Selecting Values of Input Variables in the Analysis of Output from a Computer Code." *Technometrics* 21(2):239-245.

Moskowitz, H., and R. Sarin. 1983. "Improving the Consistency of Conditional Probability Assessments for Forecasting and Decision Making." *Management Science* 29(6):735-749.

Paté-Cornell, E. 1984. "Fault Trees vs. Event Trees in Reliability Analysis." *Risk Analysis* 4(3):177-186.

Stein, M. 1987. "Large Sample Properties of Simulation Using Latin Hypercube Sampling." *Technometrics* 29(2):143-151.

U.S. Nuclear Regulatory Commission. 1975. *Reactor Safety Study: Assessment of Accident Risk in U.S. Commercial Nuclear Plants.* WASH-1400, NUREG-75/014. Washington, D.C.

U.S. Nuclear Regulatory Commission. 1991. *Severe Accident Risks: An Assessment for Five U.S. Nuclear Power Plants*, NUREG-1150, January. Available at www.nrc.gov/reading-rm/doc-collections/nuregs/staff/. Accessed February 22, 2007.

U.S. Nuclear Regulatory Commission. 1994. *A Review of NRC Staff Uses of Probabilistic Risk Assessment*, NUREG-1489, March. Washington, D.C.

The White House. 2007. Homeland Security Presidential Directive 18 [HSPD-18]: *Medical Countermeasures Against Weapons of Mass Destruction.* Available at www.fas.org/irp/offdocs/nspd/hspd-18.html. Accessed January 16, 2008.

4

Department of Homeland Security Decision Requirements for Risk Management

With finite resources for biodefense, the United States must decide how to invest optimally to best mitigate bioterrorism risk. Reducing uncertainty in risk analysis results has no direct impact on risk reduction; only the implementation of effective risk management strategies can reduce risk.

—Department of Homeland Security, *Bioterrorism Risk Assessment*, 2006

RISK MANAGEMENT REQUIRES TIMELY, ACCURATE INFORMATION

Those who, for reasons enumerated in previous chapters, have need of information from the Department of Homeland Security (DHS) regarding the risk of terrorist acts need that information to be accurate, timely, and valid. In 2002, the General Accounting Office (now Government Accountability Office) described the need to acquire and use the following information about the risk of terrorist acts in order to achieve homeland security goals and objectives: (1) Who will do what and for what reason? (2) How (in what form), where, and when will they do it? and (3) What will they use in order to do it? (GAO, 2002). Additionally, for bioterrorism, stakeholders and decision makers need answers to several types of questions (Danzig, 2003; Fischhoff et al., 2006; Whitworth, 2006):

- What is the probability of a particular biological terrorist threat and how imminent is an event? What would the consequences of the event be, characterized in terms of when, where, and in what populations?
- Can real-time detection be achieved to determine if an event has already occurred, if it is part of a larger plan, or if it is a false alarm (perhaps intentionally generated by terrorist actions)? How many ill and/or affected people would be expected for attacks by different agents? How fast would different infectious agents in a bioterrorist attack spread? How fast would people die from different agents? How sick would survivors be?
- What are the most effective ways to manage that risk? How effective and feasible are different strategies for the prevention, containment, and reduction of consequences of a bioterrorism event? Would public health measures be able to limit the spread of the infection to a small proportion of the total population? What impact would the availability and delivery of effective interventions have on potential consequences? How much lead time (i.e., response time) is needed to implement effective interventions? How much lead time will surveillance provide? What assets should be pre-positioned where and when to reduce and or manage the risk? Are transportation and communication resources sufficient to handle the surge in usage?
- How many and what type of staff would be needed to prevent, respond to, contain, or manage the consequences of a bioterrorist attack? How should clinics be designed for optimal provision of services in a crisis (e.g., mass vaccination, drug dispensing, patient triage, flow, and care) to effectively minimize the consequences of a biological terrorism attack?

These questions lead to further questions. For example, what specific interventions would be needed (i.e., vaccination for smallpox versus antibiotics for anthrax)? How fast should they be delivered? What staff and supplies will need prepositioning? (Cooper, 2006).

In reviews of unsuccessful past attempts to prevent, prepare for, or respond efficiently to terrorist attacks, the inability to put the right information into the hands of key decision makers at the right time and in an understandable format has been identified as a major factor in those failures. Failure can result in increased suffering and fatalities through setting the wrong priorities and policies, in the underfunding of important programs, and in poorly conceived plans (Aaby et al., 2006). Several barriers to the effective use of information in decision making for the prevention of, preparedness for, and response to bioterrorism have been identified (Ware et al., 2002):

- There are many stakeholders with varying degrees of authority for making and implementing decisions (e.g., the responsibilities for domestic bioterrorist attacks in the United States may involve more than 100 different government organizations).

- Authorities, roles, responsibilities, tasks, and indicators of success frequently are unclear and poorly understood or coordinated.
- High volumes of different types of data and information (e.g., subjective judgments, objective observations, historical data, analytical data, probabilistic data, modeling, simulation results) from disparate sources are presented in nonstandard and often poorly understood formats, flooding the system as crises are unfolding.
- Significant organizational friction frequently exists among the producers, owners, stakeholders, and consumers of information. Critical information is frequently owned or held by public and/or private organizations, accompanied by a general reluctance to share information across these sectors.
- The producers and owners of important information often compartmentalize it in order to protect security or to protect sources, at the expense of the timely and integrated sharing of data and interpretation of information.
- Decision makers often have widely varying objectives and frequently little understanding of the medical and scientific background needed to inform their decisions.

THE BIOLOGICAL THREAT RISK ASSESSMENT SHOULD SUPPORT RISK MANAGEMENT

The GAO (2002) report identified risk assessment as an important tool and source of information for strategic decision making for the prevention of bioterrorism, risk reduction, preparedness, and response to bioterrorism. As described in Chapter 3 of this report, the DHS Biological Threat Risk Assessment (BTRA) of 2006 is one of the first terrorism risk assessment efforts to integrate information from a variety of sources to meet information needs. Further, the BTRA of 2006 presents sensitivity analyses that permit an examination of the impact of different measures that could be taken to mitigate identified consequences of interest (i.e., morbidity and mortality). However, as noted in *Bioterrorism Risk Assessment*, the DHS 2006 report that describes the BTRA methodology, with finite resources for biodefense, the United States must decide how to invest optimally to best mitigate bioterrorism risk. As that report points out, risk assessment alone has no direct impact on risk reduction; "only the implementation of effective risk management strategies can reduce risk" (DHS, 2006).

Recommendation: Subsequent revision of the BTRA should increase emphasis on risk management. An increased focus on risk management will allow the BTRA to better support the risk-informed decisions that homeland security stakeholders are required to make.

The BTRA provides information to many stakeholders. DHS identifies the primary customers for information from the BTRA as follows:[1]

- *White House Homeland Security Council:* relative risks and overall vulnerabilities;
- *Department of Health and Human Services:* medical countermeasures needs;
- *Department of Homeland Security/Infrastructure Protection:* relative risks of different attack scenarios;
- *Department of Homeland Security/Office of Intelligence and Analysis:* high-leverage intelligence needs;
- *Department of Homeland Security/Science and Technology:* high-leverage scientific gaps;
- *Departments of Agriculture and of Health and Human Services:* food security;
- *Environmental Protection Agency:* water security; and
- *Department of Agriculture:* agricultural agents and protection of the food supply.

DHS stakeholders need risk analysis, including risk management, for strategic planning, operations, and forensics. Further, Homeland Security Presidential Directive 10 (HSPD-10): *Biodefense for the 21st Century* (The White House, 2004) states that the "United States requires a continuous, formal process for conducting routine capabilities assessments to guide prioritization of our on-going investments in biodefense-related research, development, planning, and preparedness. These assessments will be tailored to meet the requirements in each of these areas. Second, the United States requires a periodic senior-level policy net assessment that evaluates progress in implementing this policy, identifies continuing gaps or vulnerabilities in our biodefense posture, and makes recommendations for re-balancing and refining investments among the pillars of overall defense policy." To the extent that the BTRA of 2006 (with its subsequent improvements and revisions) is used for risk analysis, the committee believes that it is most applicable to supporting strategic decisions (those that address the setting of priorities and policies, the acquisition and pre-positioning and/or allocation of resources, and the development of infrastructure), but that it is not designed to support operations or forensics.

In 1997, the Presidential/Congressional Commission on Risk Assessment and Risk Management (1997 a,b; Omenn, 2003) agreed on a framework for environmental health risk management, which is applicable to managing risks involved with bioterrorism. This framework has six stages: (1) formulate the problem in a broad public health context, (2) analyze

[1]Rear Admiral Jay Cohen, Undersecretary of Science and Technology, Department of Homeland Security. 2007. "DHS Science and Technology: Enabling Technology to Protect the Nation." Briefing to the committee, February 9, 2007, Washington, D.C.

the risks, (3) define the options to address the risks, (4) make sound risk reduction decisions, (5) implement those actions, and (6) later evaluate the effectiveness of the actions taken.

The BTRA of 2006 focused on risk assessment, which addresses stages 1 and 2 above. However, DHS intends for its BTRA to be used by risk managers to "test and evaluate risk mitigation strategies and their impact on bioterrorism risk" (DHS, 2006, Ch. 1, p. 3). To broaden the focus to include risk management, stages 3 and 4 must be addressed. Stage 3, defining the options to address the risks, includes identifying potential countermeasures and estimating the costs of deploying the countermeasures. So that the risk assessments provided by the BTRA can be effectively used, each potential countermeasure must be mapped to a set of parameters within the model. Stage 4, making sound risk reduction decisions, requires several related activities (Boardman et al., 2006):

- *Estimation of risk reduction* (in, for example, expected lives, life-years, or quality-adjusted life-years) obtained by allocating countermeasures, as discussed in Chapter 2 of this report;
- *Optimization of the allocations,* which identifies, for each given resource level or budget, the allocation of countermeasures that maximizes total risk reduction (or equivalently, for a given level of risk reduction, identifies the least-cost deployment of countermeasures for achieving a particular level of risk), as discussed in Chapter 7 of this report;
- *Optimization of risk-benefit,* which, given the optimal allocation of resources for different budget levels and a willingness-to-pay value for incremental risk reduction, identifies the best overall level of resources (and corresponding best allocation); and
- *Valuation of options,* or consideration of how the results from the previous stages are likely to change if additional countermeasures are added, and using this information, making decisions about which additional countermeasures are most worth developing.

TRANSPARENCY OF RISK ASSESSMENT IS NECESSARY FOR SUCCESSFUL RISK MANAGEMENT

Transparency, as described by Oliver (2004), has been defined in different dictionaries as "free from guile," "candid or open," or "forthright," and has been applied to business and organizations as "allowing others to see the truth without trying to hide or shade the meaning or altering the facts to put things in a better light." Oliver summarizes the current use of the word transparency as "letting the truth be available for others to see if they so choose, or perhaps think to look, or have the time, means, and skills to look," and involving "active disclosure."

Whether and how often risk assessment models are used in risk management will depend on the level of confidence that stakeholders have in the model's methods, the validity of assumptions and data used to develop the model, and the level of understanding of the model's outputs. Lack of confidence can be caused by an insufficient understanding of or disagreement with relationships hypothesized among variables, the mathematical foundations of the model, and/or the validity of assumptions and values assigned to the model's parameters. All of these problems can be mitigated by improved transparency.

RISK ASSESSMENT TRANSPARENCY IMPROVES CONFIDENCE

In contemplating a complex problem involving uncertainty and risk, such as that involved with the threat of bioterrorism, mental arithmetic can be riddled with error, while risk assessment models enable repeatable calculations, including sensitivity analyses, which explore the effects of uncertain parameters on important consequences. However, risk assessment models can fail to include important knowledge that is not readily quantified and/or understood, potentially compromising the validity of model outputs (Fischhoff et al., 2006).

The accuracy of quantitative bioterrorism risk assessment models and the confidence placed in them depend on the validity of the assumptions and the availability of sound data for each of the biological agents being analyzed. A good model based on strong data, such as the one reported in Whitworth (2006) that describes the difference between a response for an anthrax attack and that for a smallpox attack, can inform judgment about the effectiveness of different interventions (i.e., antibiotics for anthrax versus vaccinations for smallpox) and the pre-positioning of staff and supplies to respond to an attack effectively.

Conversely, the lack of data and/or uncertainty in model parameters also can have important implications for the degree of confidence placed in the results of risk assessment models (Elderd et al., 2006). Unfortunately, data for key parameters of many biothreat agents of concern are not available.[2] If decision makers understand and trust the model, they will be more likely to use it with different assumptions and to test different response strategies (Fischhoff et al., 2006).

While transparency is a major factor in establishing confidence and trust in the methods of and outputs from risk assessment models, modeling is an important step toward transparency as it requires that assumptions be made explicit. However, achieving transparency also requires the careful, explicit documentation of a model's mathematical and structural foundations and of the sources of data used in the analysis—a prerequisite for any scientific study—for

[2]Marc Lipsitch, Harvard School of Public Health. "Notes to the National Research Council Committee on Methodological Improvements to the DHS's Biological Agent Risk Analysis." Written communication to the committee, February 2007.

the purpose of facilitating external review. It is essential that analysts document the following: (1) how they construct risk assessment models, (2) what assumptions are made to characterize relationships among variables and parameters and the justifications for these, (3) the mathematical foundations of the analysis, (4) the source of values assigned to parameters for which there are no available data, and (5) the anticipated impact of uncertainty for assumptions and parameters (Brisson and Edmunds, 2006).

When working with classified or sensitive information, as is the case with bioterrorism prevention, preparedness, and response, there may be need to restrict the access to some information to certain groups of users to protect overall security. However, security or confidentiality concerns that can negatively affect the level of transparency reached in risk assessment modeling include the following:

- The compartmentalization of model development, algorithms, and execution for security concerns and to protect information and data sources;
- Private-sector reluctance to share information, commonly due to the protection of proprietary considerations, across sectors; and
- The need to balance civil liberties of citizens against the need to keep important classified and sensitive information out of the hands of terrorists.

Given the importance of establishing the confidence of decision makers and stakeholders in risk assessment models, it is essential to strive for the highest level of transparency possible while being sensitive to the need to restrict access to those with a need to know.

THERE ARE SEVERAL OTHER WAYS TO BUILD CONFIDENCE

The confidence of decision makers in the information generated by a risk assessment model can be increased in several ways, and increased confidence will heighten the likelihood that the information will be used. Decision makers can be engaged iteratively during model development; this is critical in order to increase their understanding of how the model is being constructed and to ensure that their information needs will be met. Complex systems can be simplified to more-readily-understood scenarios and coupled with the capability of conducting real-time sensitivity analyses. In addition, experts can conduct independent, periodic external reviews; doing so is critical in order to assure stakeholders that the appropriate inputs and models are being used for risk assessment. If the risk assessment model and/or its assumptions are not accurate or appropriate, the results of a model and accompanying sensitivity analyses can give a false sense of security in the results, may lead to inappropriate policy decisions (Brisson and Edmunds, 2006), and ultimately will lead to a lack of confidence in the use of these models.

THE DEPARTMENT OF HOMELAND SECURITY'S BTRA OF 2006 WAS NOT TRANSPARENT

As described in Chapter 3 of this report, the model used in the BTRA of 2006 is extremely complex, with 17 stages and thousands of parameters for each of 28 biothreat agents of concern (DHS, 2006). The considerable data that are lacking for many of the parameters and probabilities in the model may lead to questions about the validity of the model's output and to a lack of confidence and trust in the results. Moreover, the results of simulations are presented in graphs, charts, and tables that are also complex and difficult to interpret and use.

The committee also finds the documentation for the model used in the BTRA of 2006 to be incomplete, uneven, and extremely difficult to understand. The BTRA of 2006 was done in a short time frame. However, deficiencies in documentation, in addition to missing data for key parameters, would make reproducing the results of the model impossible for independent scientific analysis. For example, although Latin Hypercube Sampling is mentioned in the description of the model many times as a key feature, no actual sample design is specified. Although antithetic sampling methods (e.g., matched samples or reused random number streams) evidently are employed, insufficient details are provided on how or where these numbers are generated, precluding a third party, with suitable software and expertise, from reproducing the results—violating a basic principle of the scientific method.

Finally, the current BTRA implementation must be run on a custom computer cluster in a DHS contractor facility taking many hours to compute; also the data are cumbersome to prepare. This makes answering "what-if" questions of the type that stakeholders are likely to ask an expensive and slow process.

During the course of this study, in response to technical questions posed by three members of the committee, DHS provided the committee with a technical document (DHS, 2007) that includes answers to the questions posed and which became an essential piece of documentation missing from the original publications provided to the committee. In addition, the committee asked another expert[3] to make an independent review of the methodology employed for the model used in the BTRA of 2006 using the originally published material and the technical addendum. The independent review is reproduced in this report as Appendix I. The author of that review encountered difficulties similar to those described here; one of three suggestions in Appendix I is "to report future results in a scientific fashion than can be reviewed by scientists"—a suggestion that is echoed in the next recommendation of this committee (see below).

Some of the probabilities of important consequences in the model used in the BTRA of 2006 were extremely close

[3]Alan R. Washburn, Distinguished Professor Emeritus of Operations Research, Naval Postgraduate School, Monterey, California.

to zero. When risk is expressed in numerical form, whether or not decision makers are motivated to take action will frequently depend on how confident they are in the number. Human beings are known to have difficulty in rationally processing numbers and probabilities (Paulos, 1988; Tversky and Kahneman, 1973, 1974), and when probabilities are extremely small, decision makers will often give them greater weight than is appropriate or possibly ignore them altogether, at great peril when there are potentially significant and large consequences. Thus, extreme caution is needed to avoid an under- or overinterpretation of results that may cause errors in decision making when probabilities of consequences are estimated to be near zero, such as is the case with the model used in the BTRA of 2006. Systematic use of well-grounded models can guide decision makers to account for these probabilities correctly.

Recommendation: DHS should maintain a high level of transparency in risk assessment models, including a comprehensive, clear mathematical document and a complete description of the sources of all input data. The documentation should be sufficient for scientific peer review.

To carry out this recommendation, DHS should do the following:

- Solicit input from multiple stakeholders involved with the prevention of bioterrorism (whether directed at humans or at agriculture), preparedness, and response throughout the development of the model in order to enhance their understanding of and therefore their confidence in the model, its data inputs, and its results;
- Clearly state the objectives and carefully define the input variables, sources of data, and associated consequence models; make assumptions explicit; and justify the values that are assigned to variables, parameters, and probabilities;
- Provide a guide to facilitate the interpretation of results, especially in the context of important outcomes that are estimated to occur with probabilities approaching zero; and
- Conduct a scientific, periodic external review of the validity of the risk assessment model's development and analysis; carefully and completely document how the model is developed and its mathematical foundations, using terms from a widely accepted, standard technical lexicon in understandable language, such that an independent, external panel of experts can duplicate the results; have an independent blue team perform complete scenario dissection for selected paths through the entire event tree; and take care to allow the widest possible review subject to security requirements.

THE BTRA SHOULD BECOME A DECISION SUPPORT SYSTEM

Decision support systems (DSSs) are interactive information technology platforms that facilitate the use of information in complex decision making. The goals of DSSs are to improve the efficiency with which users make decisions and to improve the effectiveness of their decisions (Shim et al., 2002; Pearson and Shim, 1995). DSSs are especially helpful in decision-making situations where there are multiple decision makers with different roles, functions, and responsibilities, and different types and sources of data and databases.

There are many different designs for DSSs, but in general they include the following components: (1) database management capabilities that provide access to relational databases, information, and knowledge from a variety of sources; (2) modeling and modeling management functions; and (3) a simple user-interface component that supports interactive queries, reporting, and graphic functions (Marakas, 1999; Druzdzel and Flynn, 2002; Shim et al., 2002; Ware et al., 2002). DSSs have been developed to support specific decisions involved with bioterrorism prevention, preparedness, and response (Bravata et al., 2002, 2004; Ware et al., 2002). Bravata et al. (2002) identified 217 information technology DSSs that were of potential use to clinicians and public health officials in the event of a bioterrorism event. One example of a DSS that facilitates data-based decision making for resource allocation problems and emergency response planning is a stand-alone, large-scale DSS, called RealOpt. RealOpt pairs a flexible simulation component with a set of analytical, decision-making algorithms. The system comprises three integrating components:

- A simulation manager that runs simulations with changes in input parameters in order to investigate behavior and bottlenecks in the system for different scenarios, and calculates various outcome statistics (e.g., average wait time, queue length, and utilization rates);
- An optimization manager that stores algorithms and fast heuristics and iteratively calls on the simulation manager to resolve and update resource-allocation statistics (e.g., to maximize throughput, to minimize staff usage to satisfy a specified throughput); and
- A user-interface manager and linker module that connects the input of data to a display of results, including a graphics algorithm that allows users to design specific floorplans of different patient care and dispensing facilities for vaccinations and different medications (Lee et al., 2006).

The characteristics of DSSs that are effective in giving decision makers access to the understandable information that they need and can use for decision making are as follows:

- The DSS clearly states its objectives and desired outcomes (i.e., timely, quality decisions);
- It addresses consequence and identifies key questions of stakeholders who were involved in framing the problems;
- It is user-friendly for a variety of stakeholders and does not require sophisticated information technology skills for its operation;
- The DSS is flexible, efficient, and includes an easy-to-access help desk and documentation;
- It is portable across different computer platforms and personal digital assistants;
- It provides results that are well matched to decision objectives. Decision makers can ask for and easily get results of new simulations reflecting different assumptions on how an event will present alternative responses and interventions leading to different outcomes;
- The DSS requires minimal computation time for simulation—seconds or minutes versus hours for individual simulation runs;
- It provides accurate results and information that can be used to gauge how much confidence can be placed in model outputs. Systematic checks of data quality are built in to the analysis system and display of results;
- It has displays that are simply designed with high-resolution data; it has relevant information presented and conveyed in an understandable way and accessed easily; and its tables and graphs are well labeled. The DSS's displays should be accompanied by annotation, details, other supplements, including limitations, aids to interpretation of risk, confidence limits around risk, and confidence in solutions.

Recommendation: Subsequent revision of the BTRA should enable a decision support system that can be run quickly to test the implications of new assumptions and new data and provide insights to decision makers and stakeholders to support risk-informed decision making.

Use Scenarios

A successful DSS, as described above, would facilitate the use of scenarios. Mathematical models (e.g., risk assessment models) often are so complex that their results are not easily understood, met with confidence, and used. Decision makers commonly deal with the uncertainty of future events by using "what-if" scenarios, which can bound uncertainty and bring multiple stakeholders together to consider a shared, selected set of hypothesized chains of events in narrative form, and to consider alternatives (Pomerol, 2001). Scenarios can make abstract or nebulous threats more concrete, which can help decision makers avoid becoming lost in trying to assimilate large numbers of variables, relationships, and parameters with extremely small probabilities. Scenarios are especially useful when decision makers are inexperienced in systems thinking. They may help inexperienced systems thinkers avoid using unrealistic assumptions, which can lead to the development of incomplete, infeasible, or ineffective plans (Whitworth, 2006).

Danzig (2003) described a number of situations in which planning scenarios could be used. First, they can bring awareness to sets of specific circumstances and hypothesized chains of events, which, if understandable and conveyed in a compelling manner such that the decision maker has confidence in the method, will have a greater likelihood for resulting in action. Second, they can help in the development of coordinated actions and plans among multiple stakeholders by keeping everyone focused on the same narrative or alternative. Third, a planning scenario can serve as a reference case against which alternative strategies can be compared and tested. Fourth, planning scenarios can be used to establish resource and other requirements needed to prevent or respond to potential events.

Critics of scenarios are concerned that their use may make assumptions unclear or inexplicit, complicating external review and assessment and making their validity difficult to assess. When insufficient detail is provided, different stakeholders may arrive at different perceptions of a scenario and end up coming to incompatible conclusions or developing uncoordinated or incompatible plans. For this reason, scenarios must reflect the most complete, explicit, and transparent details available and allow for a ready comparison of perceptions among the various stakeholders. Finally, scenarios can become too rigid. They require continual updating as new information becomes available. However, effectively planning for these possibilities can mitigate these organization problems. The committee recognizes the difficulty in preparing and validating accurate and useful scenarios. For that reason, it suggests that, as with other BTRA documentation, any such scenarios be peer reviewed.

Sensitivity Analysis Is Important for Validation

Sensitivity analysis has been defined as the determination of how "uncertainty in the output of a model (numerical or otherwise) can be apportioned to different sources of uncertainty in the model input" (Saltelli and Tarantola, 2002). The purposes of sensitivity analyses are to (1) give users of risk assessment models information that they can use to identify key parameters and explore a range of impacts that can be expected with changes in input and parameter values, and to evaluate the confidence they can place on model outputs; (2) identify sources of uncertainty in the model when assumptions and parameters vary across possible scenarios; (3) aid planners in comparing alternative strategies and test how a given plan would work should assumptions be wrong;

(4) help decision makers make the best possible decisions in the presence of uncertainty; and (5) set priorities for the collection of additional information (Meltzer, 2001; Whitworth, 2006).

Because considerable data are lacking for many of the parameters and probabilities in the model used for the BTRA of 2006, there may be an accompanying lack of confidence in the results. Thus, being able to conduct a sensitivity analysis is an essential feature of the BTRA model if it is to be used for decision making. Although the BTRA was apparently developed so that the impact of different parameters on consequences of importance could be assessed through sensitivity analysis, the process for running sensitivity analyses currently is not interactive and appears quite "user-*unfriendly*" and cumbersome. To see results of the model for different scenarios (changing values for parameters and in different branches of the tree for different agents), changes in values must be submitted to analysts who rerun the model. Results are then evidently available hours to days later, making the sensitivity analysis process difficult for decision makers to use immediately. Finally, the results of simulations are presented in graphs, charts, and tables that are complex and difficult to interpret and use.

For these reasons, the committee questions whether the results of the 2006 Biological Threat Risk Assessment model are answering the highest-priority questions of different decision makers; whether they are being conveyed in the most understandable, useful, and compelling manner possible; and whether the current sensitivity analysis feature is meeting information needs. User-friendly sensitivity analysis could also be a part of any DSS.

The uncertainty or lack of data and/or errors in measurement for many key variables and parameters in risk assessment models for potential bioterrorism events can affect the confidence that decision makers place on the output of the model. Accepted good modeling practices require that models be continually tested and validated by evaluating the effect of uncertainties with regard to values of parameters and probabilities of the model. Sensitivity analysis has become an accepted and important approach to the testing and validation of risk assessment models of complex systems (Borgonovo, 2006).

In the future, it will be important to move the sensitivity analysis from questions about risk assessment (for example, How does uncertainty about the infectious dose for this agent change my expected consequences?) to questions about risk management (for example, If I had improved knowledge about the infectious dose for this agent, would I adopt different countermeasure strategies?). Currently, simulation runs take an extended period of time to run owing to the complexity and size of the model and its input data; outputs from the model are not presented in easily used, interactive, understandable and compelling formats. Strategies for reducing the complexity of the model are presented in Chapter 7; strategies for tools that are easier for users to employ are presented in the next subsection.

The purpose of the sensitivity analysis will affect the method to use for the analysis. Borgonovo (2006) described three families of analytic techniques (i.e., variance-based, input-output correlation, and moment-independent analyses), the choice of which would depend on the stated purpose of the sensitivity analysis.

Precautions must be taken when drawing conclusions from sensitivity analyses, as the accuracy of the information is conditional on the validity of the underlying model structure and the methods used to exercise it. If the structure of the risk assessment model and/or the assumptions used in the model are not accurate or appropriate, then the results of a sensitivity analysis can give a false sense of security in the results and may lead to inappropriate policy decisions (Brisson and Edmunds, 2006). Extreme caution also is needed to avoid a misinterpretation or overinterpretation of results and the making of errors in decision making when valid data for parameters or probabilities are lacking.

Create a Context for Use

In addition to the approaches discussed so far, strengthening the overall environment for data-based decision making is critical. A comprehensive and continually updated set of guidelines, protocols, and checklists that provide essential details on clear courses of actions that decision makers would make, conditional on the information made available to them from risk assessment models analyzing a set of structured scenarios, must be developed, tested, and in place. These materials should be prepared for different stakeholders and should include a range of possible decisions and actions, by scenario, for different authorities, roles and responsibilities, desired outcomes, and benchmarks for tracking progress for different scenarios.

Different strategies and protocols should be practiced as tabletop and TOPOFF[4] exercises to identify areas where additional attention, planning, resource acquisition and allocation, and practice are required. A trained workforce within and across multiple sectors, agencies, and institutions in the public and private sectors is essential.

Relevant, accurate, timely information that is available in understandable formats and terms, with guides to the interpretation of outcomes, is critical. Environments that support the use of data in decision making are those in which the right resources (e.g., staff, drugs, vaccines, respirators, other) are pre-positioned strategically and available to the right staff at the right time.

[4]In Chapter 7, the committee discusses the benefits of red teaming in TOPOFF (Top Officials) exercises.

REFERENCES

Aaby, K., J.W. Herrmann, C.S. Jordan, M. Treadwell, and K. Wood. 2006. "Montgomery County's Public Health Service Uses Operations Research to Plan Emergency Mass Dispensing and Vaccination Clinics." *Interfaces* 36(6):569-579.

Boardman, A.E., D.H. Greenberg, A.R. Vining, and D.L.Weimer. 2006. *Cost-Benefit Analysis: Concepts and Practice,* 3rd Edition. Upper Saddle River, N.J.: Pearson Prentice Hall.

Borgonovo, E. 2006. "Measuring Uncertainty Importance: Investigation and Comparison of Alternative Approaches." *Risk Analysis* 20(5):1349-1361.

Bravata, D.M., K. McDonald, and D.K. Owens. 2002. *Bioterrorism Preparedness and Response: Use of Information Technologies and Decision Support Systems.* University of California, San Francisco—Stanford Evidence-based Practice Center. AHRQ Publication No. 02-E028. Available at www.ahrq.gov/clinic/tp/bioittp.htm. Accessed January 30, 2007.

Bravata, D.M., V. Sundaram, K.M. McDonald, W.M. Smith, H. Szeto, M.D. Schleinitz, and D.K. Owens. 2004. "Evaluating Detection and Diagnostic Decision Support Systems for Bioterrorism Response." *Emerging Infectious Diseases* 10(1):100-108.

Brisson, M., and W.J. Edmunds. 2006. "Impact of Model, Methodological, Parameter Uncertainty in the Economic Analysis of Vaccination Programs." *Medical Decision Making* 26(5):434-446.

Cooper, B. 2006. "Poxy Models and Rash Decisions." *Proceedings of the National Academy of Sciences of the United States of America* 103(33):12221-12222.

Danzig, R. 2003. *Catastrophic Bioterrorism—What Is to Be Done.* Center for Technology and National Security Policy. Washington D.C.: National Defense University.

DHS (Department of Homeland Security). 2006. *Bioterrorism Risk Assessment.* Biological Threat Characterization Center of the National Biodefense Analysis and Countermeasures Center. Fort Detrick, Md.

DHS. 2007. "A Lexicon of Risk Terminology and Methodological Description of the DHS Bioterrorism Risk Assessment." Written communication to the Committee on Methodological Improvements to the Department of Homeland Security's Biological Agent Risk Analysis. April 14, 2007.

Druzdzel, M., and R.R. Flynn. 2002. "Decision Support Systems." *Encyclopedia of Library and Information Science,* 2nd Edition. Allen Kent (ed.), New York: Marcel Dekker. Available at www.pitt.edu/~druzdzel/psfiles/dss.pdf. Accessed January 30, 2008.

Elderd, B.D., V.M. Dukic, and G. Dwyer. 2006. "Uncertainty in Predictions of Disease Spread and Public Health Responses to Bioterrorism and Emerging Diseases." *Proceedings of the National Academy of Sciences of the United States of America* 103(42):15693-15697.

Fischhoff, B., W.B. Bruin, U. Güvenç, D. Caruso, and L. Brilliant. 2006. "Analyzing Disaster Risks and Plans: An Avian Flu Example." *Journal of Risk and Uncertainty* 33(1):131-149.

GAO (General Accounting Office). 2002. *National Preparedness: Integrating New and Existing Technology and Information Sharing into an Effective Homeland Security Strategy.* GAO-02-811T. Available at www.gao.gov/new.items/d02811t.pdf. Accessed January 30, 2008.

Lee, E.K., S. Maheshwary, J. Mason, and W. Glisson. 2006. "Large-Scale Dispensing for Emergency Response to Bioterrorism and Infectious-Disease Outbreak." *Interfaces* 36(6):591-607.

Marakas, G.M. 1999. *Decision Support Systems in the Twenty-First Century.* Upper Saddle River, N.J.: Prentice Hall.

Meltzer, D. 2001. "Addressing Uncertainty in Medical Cost-effectiveness Analysis. Implications of Expected Utility Maximization for Methods to Perform Sensitivity Analyses and Use of Cost-effectiveness Analysis to Set Priorities for Medical Research." *Journal of Health Economics* 20(1):109-129.

Oliver, R.W. 2004. *What Is Transparency?* New York: McGraw-Hill.

Omenn, G.S. 2003. "On the Significance of 'the Red Book' in the Evolution of Risk Assessment and Risk Management." *Human and Ecological Risk Assessment* 9(5):1155-1167.

Paulos, J.A. 1988. *Innumeracy: Mathematical Illiteracy and Its Consequences.* New York: Hill and Wang.

Pearson, J.M., and J.P. Shim. 1995. "An Empirical Investigation into DSS Structures and Environments." *Decision Support Systems* 13(2):141-158.

Pomerol, J.C. 2001. "Scenario Development and Practical Decision Making Under Uncertainty." *Decision Support Systems* 31(2):197-204.

Presidential/Congressional Commission on Risk Assessment and Risk Management. 1997a. *Framework for Environmental Health Risk Management.* Final Report Vol. 1. Washington D.C.: U.S. Government Printing Office. Available at www.riskworld.com/Nreports/1997/risk-rpt/pdf/EPAJAN.pdf. Accessed January 30, 2008.

Presidential/Congressional Commission on Risk Assessment and Risk Management. 1997b. *Risk Assessment and Risk Management in Regulatory Decision Making.* Final Report Vol. 2. Washington D.C.: U.S. Government Printing Office. Available at www.riskworld.com/Nreports/1997/risk-rpt/volume2/pdf/v2epa.pdf. Accessed January 30, 2008.

Saltelli, A., and S. Tarantola. 2002. "On the Relative Importance of Input Factors in Mathematical Models: Safety Assessment for Nuclear Waste Disposal." *Journal of the American Statistical Association* 97(459):702-709.

Shim, J.P., M. Warkentin, J.F. Courtney, D.J. Power, R. Sharda, and C. Carlson. 2002. "Past, Present, and Future of Decision Support Technology." *Decision Support Systems* 33(2): 111-126.

Tversky, A., and D. Kahneman. 1973. "Availability: A Heuristic for Judging Frequency and Probability." *Cognitive Psychology* 5(2):207-232.

Tversky, A., and D. Kahneman. 1974. "Judgment Under Uncertainty: Heuristics and Biases." *Science* 185(4157):1124-1131.

Ware, B.S., A. Beverina, L. Gong, and B. Colder. 2002. *A Risk-Based Decision Support System for Antiterrorism.* Available at www.dsbox.com/Documents/MSS_A_Risk-Based_Decision_Support_System_for_Antiterrorism.pdf. Accessed January 30, 2007.

The White House. 2004. Homeland Security Presidential Directive 10 [HSPD-10]: *Biodefense for the 21st Century.* Available at www.fas.org/irp/offdocs/nspd/hspd-10.html. Accessed January 16, 2008.

Whitworth, M.H. 2006. "Designing the Response to an Anthrax Attack." *Interfaces* 36(6):562-568.

5

Risk Assessment for Unknown and Engineered Biothreat Agents

How do we avoid becoming beguiled by the risks we have already experienced, and distracted from those that our enemy might be planning in the future?

—Department of Homeland Security Secretary Michael Chertoff at Homeland Security Policy Institute, March 16, 2005

BIOLOGICAL THREAT RISK ASSESSMENTS NEED TO INCLUDE UNKNOWN AND ENGINEERED AGENTS

Most of this report deals with assessing bioterrorism risk and prioritizing risks associated with known biological agents. However necessary the focus on risk associated with *known* biological agents is, the committee strongly believes that it is not sufficient. The Department of Homeland Security's (DHS's) Biological Threat Risk Assessment (BTRA) of 2006 only considers threats already known and at least partially characterized. However, the biological threat spectrum is dynamic (Petro et al., 2003; IOM and NRC, 2006) and therefore requires a proactive approach. Some agents on the Category A list[1] (such as several of the hemorrhagic fevers) of the Centers for Disease Control and Prevention (CDC) were discovered only within the past few decades, and there are undoubtedly many more pathogens still undiscovered in nature (IOM, 1992, 2003; Morse, 1991, 1995). Some of them may be similar to agents already known. Others, such as Nipah virus infection and severe acute respiratory syndrome (SARS) (discovered in 2003), may have completely unexpected characteristics. In addition, previously unknown pathogens will continue to be discovered or to evolve from nature. Some may be adopted by adversaries as "bioweapons of convenience," just as the current biothreat agents were all zoonotic diseases (animal diseases that can be transmitted to humans) adopted by older bioweapons programs because their biological or physical characteristics made them suitable.

So far, with respect to biothreat agents, nature has been the greatest source of novelty, but the rapid advances in molecular biology and biotechnology and the increasing understanding of pathogenesis (the mechanisms by which these organisms cause disease) cannot be ignored. These advances suggest that the future will be even more complex and uncertain (IOM and NRC, 2006). Agents can be modified for new properties in a variety of ways. Discoveries in this area in the past few years have included the following:

- *Poxviruses with an IL-4 gene insert that can cause severe disease in immunized or genetically resistant animals* (Jackson et al., 2001). IL-4 (interleukin-4) is a mammalian protein that serves as one of several important regulators of immune response. Ectromelia (mousepox) virus is similar in lethality and contagiousness to smallpox in humans and is closely related to the smallpox virus. As with human smallpox, mice can be protected by the same vaccine that is used to protect humans against smallpox. However, when immunized mice are infected with the IL-4 modified mousepox virus, the effects are severe and similar to those in unimmunized mice. In addition, strains of mice that normally would be genetically resistant are not resistant to the modified virus, but become sick in the same way that more susceptible mice do. Fortunately, for reasons not clearly understood, these engineered strains do not transmit well to others. However, one can anticipate that a future technically adept adversary could solve this problem.
- *Anthrax modified with a gene insert from a nonpathogenic relative that can defeat a live anthrax vaccine* (Pomerantsev et al., 1997). Inserting a particular gene from a relatively harmless anthrax relative, *Bacillus cereus,* made anthrax able to infect and kill animals (hamsters) that had been immunized with the standard live vaccine used in Russia for human protection. This live vaccine, known as STI, is generally considered highly effective, and (although comparative data are lacking) is widely thought to be equivalent in efficacy to the protective antigen (PA) protein-based vaccine used in the United States and United Kingdom. Earlier, at the 1995 International Anthrax Meeting in Salisbury, United Kingdom, the same Russian group had reported

[1]For the CDC list, see www.bt.cdc.gov/agent/agentlist-category.asp#a. Accessed February 25, 2008.

developing multi-drug-resistant anthrax. Although a vaccine strain was used for this experiment, it could just as easily have been done with virulent anthrax.

- *Reconstruction of viable 1918 pandemic influenza virus* (Tumpey et al., 2005). The virus responsible for the most notorious influenza pandemic in recorded history (with an estimated 50 million human deaths worldwide) was recently reconstructed from several different sources using molecular techniques. This tour-de-force of molecular biology (by Jeffery Taubenberger and colleagues) made it possible to study the 1918 pandemic virus for the first time. The virus could be grown and tested in several animal species, in which it caused severe disease. The purpose of the work was constructive, to better understand how the 1918 pandemic virus caused such serious disease. Even more recently, it was shown that two specific amino acid changes in the hemagglutinin (HA) surface protein of the H5N1 avian influenza virus would enable it to bind to human, rather than avian, tissues, a necessary first step in being able to readily infect humans (Yamada et al., 2006).

The motives for all the work cited here are ostensibly benign: to better understand these dangerous pathogens. But it is easy to imagine how the same techniques could be applied to other uses. At present, conducting this work requires specialized laboratory expertise at the postgraduate level or above, and the influenza genetic system is currently beyond the technical capabilities of all but a few experts. However, advances in biotechnology will make all of these techniques more accessible in the future (IOM and NRC, 2006). The powerful molecular technique for selectively copying deoxyribonucleic acid (DNA) known as the polymerase chain reaction (PCR) was so esoteric in the early to mid-1990s that performing it required painstaking technique by experienced scientists. PCR has now become so widely used and routine that it is commonplace in high school science projects and is even taught to schoolchildren visiting museum exhibitions. As another example, the complete chemical synthesis of the poliovirus genome (a small ribonucleic acid [RNA] virus) required several years of work by experts, including overcoming a number of technical difficulties (Cello et al., 2002). Since then, the George Church Laboratory at Harvard University has devised microchips that could be used to synthesize even larger genomes with far less effort (Tian et al., 2004), and other large-scale rapid DNA synthesis methods are at the advanced development stage. There has also been academic interest in "synthetic biology," a kind of engineering using biological component parts to make entities with desired functions (Bio FAB Group et al., 2006). It is clear that future possibilities will be limited more by imagination than by technical obstacles.

Very few individuals today are capable of using these techniques, and it is likely to be some time before other than state-sponsored terrorists will be able to take advantage of such technological advances. In the meantime, conventional threats are likely to predominate. Nevertheless, if the history of PCR and other scientific advances is any indication, the use of biotechnology to engineer novel threats will come in time. It has been suggested that engineering "advanced bioweapons" is a natural extension of advancing biotechnology. In the words of the authors of a recent publication on this subject (Petro et al., 2003, p. 161):

> Advances in biological research likely will permit development of a new class of advanced biological warfare (ABW) agents engineered to elicit novel effects. . . . Such new agents and delivery systems would provide a variety of new use options, expanding the BW paradigm. Although ABW agents will not replace threats posed by traditional biological agents such as *Bacillus anthracis* (anthrax) and Variola (smallpox), they will necessitate novel approaches to counterproliferation, detection, medical countermeasures, and attribution.

In consideration of these possibilities, the White House recently released Homeland Security Presidential Directive 18 (HSPD-18): *Medical Countermeasures Against Weapons of Mass Destruction* (The White House, 2007) as a follow-up to the original biodefense strategy embodied in HSPD-10 (The White House, 2004). HSPD-10 is the document that, among other tasks, instituted the regular threat assessment that constitutes the main thrust of this committee's work. Setting out the outlines of the U.S. biodefense strategy, HSPD-10 states that "[t]he essential pillars of our national biodefense program are: Threat Awareness, Prevention and Protection, Surveillance and Detection, and Response and Recovery." HSPD-18 takes this strategy a step farther, mandating that "[o]ur Nation will use a two-tiered approach for development and acquisition of medical countermeasures, which will balance the immediate need to provide a capability to mitigate the most catastrophic of the current CBRN (chemical, biological, radiological, and nuclear) threats with long-term requirements to develop more flexible, broader spectrum countermeasures to address future threats." The biodefense tasks are divided into "Tier I: Focused Development of Agent-Specific Medical Countermeasures" for current and anticipated biological threats and "Tier II: Development of a Flexible Capability for New Medical Countermeasures" (The White House, 2007). The latter specifically recognizes the diversity of possible future biological threats, both natural and engineered, and the need for broad-spectrum solutions.

The BTRA of 2006 does not lend itself readily to the rapid assessment of new threats. Cybersecurity presents similar contrasts of comprehensiveness versus flexibility. Buckshaw et al. (2005) developed a quantitative risk model based on the adversary's attack preferences instead of the adversary's probabilities of attack. This has certain advantages (e.g., Buckshaw at al. [2005, p. 24] note, "Adversary attack preferences are easier to measure and help develop the mitigation strategy. We need to consider all attacks since a capable and adaptive threat will constantly change their actions in response to our assurance activities."). However, Buckshaw et al. (2005, p. 36) also note the same drawback of needing large amounts of reasonably accurate data: "Data are the benefit and bane. . . . If time is spent to get the data required of a quantitative risk model, then one can produce insightful and clear recommendations for the decision maker. Because the data requirements are so large, we recommend that [this] be used only on critical information systems."

INCLUDING UNKNOWN AND ENGINEERED AGENTS IS CHALLENGING BUT POSSIBLE

There are several possible ways to deal with the unpredictable and dynamic future for both natural and engineered agents:

- *Concentrate on known agents, and develop new risk assessments as each new threat is identified (e.g., from intelligence).* While this avoids speculation about future possibilities and possibly unnecessary work, it has several potential weaknesses. Risk assessment models such as the BTRA of 2006 require large amounts of specific information about the agent and its properties. Even with such well-known agents as *Bacillus anthracis* (anthrax) or *Yersinia pestis* (plague), the critical data are approximate at best, with uncertainties that have proven elusive to quantify (indeed, some data may vary by strain of organism and conditions of assay). Obtaining these data for newly recognized or newly engineered agents is likely to be even more difficult, and to exact a significant time lag.
- *Attempt to identify every potential future threat.* The committee believes that both the complexity of nature and unforeseen advances in biotechnology will make this task infeasible and may lead to a false sense of security by leaving the United States unprotected against newly engineered pathogens. There are too many theoretical possibilities and, barring reliable intelligence information, prioritization is likely to be exceedingly difficult.
- *Consider "more-generic" categorization of agents and risks into groups by various properties, identifying the most critical variables.* This would provide a general framework that could be used to classify newly identified threats as they appear on the basis of even limited information. In addition, this approach may suggest mitigation strategies that already apply to existing pathogens. The committee favors this open-ended approach for new and emerging pathogens.

What are some possible standard situations or classifications to use in an effort to anticipate future threats? Consider the analogy presented by grouping threats in information assurance analyses. McCumber (1991) introduces a qualitative model for information security that incorporates the three concepts of information characteristics, information states, and security countermeasures. DHS could seek, with its customers, multiple classification schemes that are most useful to each of its customers' end-user communities. Rather than requiring specific numbers (such as R_0, or infection rate), identifying combinations of key variables as qualitative categories would also help model vulnerabilities and prioritize concerns that have not yet been foreseen by existing analyses.

Although analyzing newly emerging threats may seem a daunting task, it actually appears to be quite feasible. While the number of all possible combinations of characteristics is enormous, it would not be necessary to deal with such vast numbers of combinations in practice, because the analysis would be limited to a number of key characteristics at relatively broad qualitative levels. The high, medium, and low threat categories appear meaningful to some users and might be sufficient.

Several efforts have been made to define the weapon or terrorist potential of microbial agents. For example, Casadevall and Pirofski (2004) recently attempted to identify the characteristics that might contribute to the weapon potential of an agent. Such efforts have repeatedly identified several of the same characteristics—for example:

- Ease of acquisition,
- Transmissibility,
- Mode of spread (person to person or by direct exposure only, or both),
- Case fatality rate,
- Ease of dissemination,
- Frequency of serious sequelae (e.g., blindness or neurological disease) in survivors, and
- Availability and efficacy of countermeasures (vaccine or other prophylactic measures) and therapeutic measures.

All of these criteria are included in the BTRA of 2006. However, a true quantitative estimate is virtually impossible for newly recognized or poorly characterized agents. The committee has prepared a rapid assessment tool that can be applied to any newly recognized agent for which there are only very limited specific data, and it suggests that DHS consider development of such a tool. Such a rapid assessment tool could use attributes similar to those listed above, with the agent qualitatively categorized as being a low, medium, or high threat with respect to each attribute; these categories could be assigned numerical scores (e.g., 1, 2, and 3) and these scores used as a signature to compare with known pathogens. Although such qualitative assessments cannot replace detailed risk assessment, the use of such a rapid assessment tool can aid DHS in focusing on areas where further expansion of possibilities and more evaluation are needed.[2]

[2]T. Cox, Cox Associates. 2002. "What's Wrong with Risk Matrices." Unpublished.

TABLE 5.1 An Illustrative Approach to Rapid Assessment, With Some Examples

Attribute	Anthrax	Brucella	Ebola	Salmonella	Smallpox
1. Ease of acquisition or synthesis	3	3	1	3	1
2. Environmental stability	3	1	1	2	2
3. Transmissibility person to person	1	1	1	2	3
4. Case fatality rate (untreated)	3	1	3	1	3
5. Ease of dissemination (estimated)	2	2	1	3	2
6. Frequency of serious sequelae (e.g., blindness or neurological disease) in survivors	2	2	2	1	3
7. Lack or unavailability of useful countermeasures or treatment[a]	1	3	3	2	2
8. Need for immediacy in diagnosis and treatment to ensure patient survival	3	1	3	1	3
Total	18	14	15	15	19

NOTE: 1 = Low threat, 2 = Medium threat, 3 = High threat.
[a]High threat = Countermeasures/therapeutics nonexistent or not readily available.

The committee's rapid assessment tool is an adaptation of the Multi-Attribute Risk Analysis (MARA) step 1 used in the development of the DHS risk assessment process as described in the DHS report *Bioterrorism Risk Assessment* (DHS, 2006, Ch. 6, pp. 1-28). In MARA, there are 28 attributes, each scored 0 through 4 by a panel of experts. Aggregating these categories to broadly reflect several key characteristics, such as those listed above, could form the basis for a rapid assessment tool. Although results would, of course, be approximate, this rapid assessment would help DHS and its partner agencies determine whether a newly identified agent is a high priority for additional consideration.

As a hypothetical illustration of this sort of rapid assessment, a template and some worked examples are shown in Table 5.1.

While engineered agents are by definition novel, an engineered agent will likely be designed for a specific function. The committee therefore anticipates that the evaluation of such agents would be similar to the evaluation of novel natural agents.

Additionally, rigorous sensitivity analyses applied to the BTRA (as recommended by the committee elsewhere in this report) can help identify the key characteristics for a rapid assessment and should be done regularly and on a variety of parameters. Since many of the parameters input to the BTRA result from the elicitation of expert opinion, the threats that emerge may only reinforce general expert consensus. Alternative consequence analyses of different routes of administration (such as large-scale food contamination) should be rigorously tested to ensure that these results are robust. It is also possible that even improvised suboptimal routes used by an adversary may cause significant morbidity and mortality, or mental health consequences. This issue is further examined in Chapter 6, "Improving Bioterrorism Consequence Assessment."

It is said that generals are adept at fighting the last war. The committee recommends expanding present approaches to form the basis of a more proactive strategy to face future threats.

Recommendation: The BTRA should be broad enough to encompass a variety of bioterrorism threats while allowing for changing situations and new information. DHS should develop a strategy for the rapid assessment of newly recognized and poorly characterized threats.

REFERENCES

Bio FAB Group, D. Baker, G. Church, J. Collins, D. Endy, J. Jacobson, J. Keasling, P. Modrich, C. Smolke, and R. Weiss. 2006. "Engineering Life: Building a Fab for Biology." *Scientific American* 294(6):44-51.

Buckshaw, D.L., G.S. Parnell, W.L. Unkenholz, D.L. Parks, J.M. Wallner, and O.S. Saydjari. 2005. "Mission Oriented Risk and Design Analysis of Critical Information Systems." *Military Operations Research* 10(2):19-38.

Casadevall, A., and L.A. Pirofski. 2004. "The Weapon Potential of a Microbe." *Trends in Microbiology* 12(6):259-263.

Cello, J., A.V. Paul, and E. Wimmer. 2002. "Chemical Synthesis of Poliovirus cDNA: Generation of Infectious Virus in the Absence of Natural Template." *Science* 297(5583):1016-1018.

DHS (Department of Homeland Security). 2006. *Bioterrorism Risk Assessment.* Biological Threat Characterization Center of the National Biodefense Analysis and Countermeasures Center. Fort Detrick, Md.

IOM (Institute of Medicine). 1992. "Emerging Infections." In *Microbial Threats to Health in the United States*, J. Lederberg, R.E. Shope, and S.C. Oaks, Jr. (eds.). Washington, D.C.: National Academy Press.

IOM. 2003. *Microbial Threats to Health: Emergence, Detection, and Response.* Mark S. Smolinski, Margaret A. Hamburg, and Joshua Lederberg (eds.). Washington, D.C.: The National Academies Press.

IOM and NRC (National Research Council). 2006. *Globalization, Biosecurity, and the Future of the Life Sciences.* Committee on Advances in Technology and the Prevention of Their Application to Next Generation Biowarfare Threats. Washington, D.C.: The National Academies Press.

Jackson, R.J., A.J. Ramsay, C.D. Christensen, S. Beaton, D.F. Hall, and A.I. Ramshaw. 2001. "Expression of Mouse Interleukin-4 by a Recombinant Ectromelia Virus Suppresses Cytolytic Lymphocyte Responses and Overcomes Genetic Resistance to Mousepox." *Journal of Virology* 75(3):1205-1210.

McCumber, J. 1991. "Information Systems Security: A Comprehensive Model." In *Proceedings: 14th National Computer Security Conference*. Baltimore, Md.: National Institute of Standards and Technology.

Morse, S.S. 1991. "Emerging Viruses: Defining the Rules for Viral Traffic." *Perspectives in Biology and Medicine* 34(3):387-409.

Morse, S.S. 1995. "Factors in the Emergence of Infectious Diseases." *Emerging Infectious Diseases* 1(1):7-15.

Petro, J.B., T.R. Plasse, and J.A. McNulty. 2003. "Biotechnology: Impact on Biological Warfare and Biodefense." *Biosecurity and Bioterrorism* 1(3):161-168.

Pomerantsev, A.P., N.A. Staritsin, Yu.V. Mockov, and L.I. Marinin. 1997. "Expression of Cereolysine AB Genes in Bacillus Anthracis Vaccine Strain Ensures Protection Against Experimental Hemolytic Anthrax Infection." *Vaccine* 15(17-18):1846-1850.

Tian, J., H. Gong, N. Sheng, X. Zhou, E. Gulari, X. Gao, and G. Church. 2004. "Accurate Multiplex Gene Synthesis from Programmable DNA Microchips." *Nature* 432(7020):1050-1054.

Tumpey, T.M., C.F. Basler, P.V. Aguilar, H. Zeng, A. Solorzano, D.E. Swayne, N.J. Cox, J.M. Katz, J.K. Taubenberger, P. Palese, and A. Garcia-Sastre. 2005. "Characterization of the Reconstructed 1918 Spanish Influenza Pandemic Virus." *Science* 310(5745):77-80.

The White House. 2004. Homeland Security Presidential Directive 10 [HSPD-10]: *Biodefense for the 21st Century*. Available at www.fas.org/irp/offdocs/nspd/hspd-10.html. Accessed January 16, 2008.

The White House. 2007. Homeland Security Presidential Directive 18 [HSPD-18]: *Medical Countermeasures Against Weapons of Mass Destruction*. Available at www.fas.org/irp/offdocs/nspd/hspd-18.html. Accessed January 16, 2008.

Yamada, S., Y. Suzuki, T. Suzuki, M.Q. Le, C.A. Nidom, Y. Sakai-Tagawa, M. Yukito, Y. Muramoto, M. Ito, M. Kiso, T. Horimoto, K. Shinya, T. Sawada, M. Kiso, T. Usui, T. Murata, Y. Lin, A. Hay, L.F. Haire, D.J. Stevens, R.J. Russell, S.J. Gamblin, J.J. Skehel, and Y. Kawaoda. 2006. "Haemagglutinin Mutations Responsible for the Binding of H5N1 Influenza A Viruses to Human-type Receptors." *Nature* 444(7117):378-382.

6

Improving Bioterrorism Consequence Assessment

. . . and the big picture is worrying about how do we protect the most people from the greatest risks most of the time.

—Department of Homeland Security Secretary Michael Chertoff
at a press conference, January 5, 2007

EXISTING KNOWLEDGE DOES NOT SUPPORT THE DETAIL IN DEPARTMENT OF HOMELAND SECURITY CONSEQUENCE MODELS

In the Department of Homeland Security's (DHS's) report on the Biological Threat Risk Assessment (BTRA) of 2006 (DHS, 2006), three measures of consequences are determined for each scenario: fatalities, illnesses, and direct economic costs.[1] These three measures are dependent on intrinsic properties of the pathogen, the details of the scenario, and the hypothesized U.S. response to the event, accounting for the effect of current U.S. medical mitigation capacity. Although in the BTRA of 2006 an analysis was conducted for the three measures of consequences, *the overall risk-informed agent prioritization is based only on mortality*. In presentations to the committee, DHS reported that it intends to take into account indirect economic costs (e.g., medical mitigation, emergency response, cleanup, and business loss) as well. Some projected improvements for future BTRAs are described in Chapter 3.

Assessing an infectious agent's impact on a population is challenging. In order to measure the health consequences, currently defined as the number of fatalities and of ill people, DHS has implemented a susceptible, exposed, infected, and recovered (SEIR) model using an off-the-shelf software package called STELLA,[2] which is run for each scenario. SEIR is a deterministic, "compartmental" model; it categorizes individuals as being in one of four compartments, representing the susceptible, the exposed, the infected, and those who have recovered. The model parameters specify the transition rates between the compartments—for example, the attack rate—as susceptible people become infected.

SEIR models can provide useful insights into the mechanics of many common infectious diseases and into the effectiveness of control strategies. However, SEIR and similar modeling approaches have limitations. Even the simplest model requires a minimum amount of parametric data: in particular, the attack rate or risk of transmission per contact, the incubation period of the disease, the number of potentially infectious contacts that a person has per unit of time, and the duration of the transmissible period. Models that assess the health consequences of the pathogens of concern for bioterrorism are difficult to parameterize owing to the lack of an adequate empirical base. Many diseases are relatively obscure and are associated with extremely limited clinical and epidemiologic data.[3] For example, for many diseases caused by agents being considered by DHS, little is known about the dose-response relationships (a major concern, as the size of a dose may determine whether symptoms occur), the duration of the incubation period, and the severity of infection. Thus, estimates of the key parameters for most bioterrorism agents must reflect a very large variance. The anthrax attacks in the United States in 2001 demonstrated low correlation between environmental exposure and infection risk. Given the great uncertainty associated with model inputs, it is important to acknowledge that generated predictions are rough approximations at best and, while useful in helping to understand a problem, should not be regarded as more than rough approximations. Exacerbating this lack of certainty in model outputs, the increasing availability of sophisticated computer software allows researchers to create highly artificial models, sometimes based on weakly defended assumptions, where the

Note: The committee thanks Jason Matheny, Ellis McKenzie, Marc Lipsitch, and Michael Boechler for reading this chapter.

[1]"Direct economic costs" here refer to the costs of hospitalization and funerals and exclude the cost of decontamination, loss of worker productivity, and so on.

[2]See www.iseesystems.com. Accessed January 30, 2008.

[3]Marc Lipsitch, Harvard School of Public Health. "Notes to the National Research Council Committee on Methodological Improvements to the DHS's Biological Agent Risk Analysis." Written communication to the committee, February 2007.

complexity and precision of results can be mistaken for accuracy.[4]

Even when data for well-studied pathogens (e.g., influenza viruses) are available, predicting the propagation of infection in a population requires understanding how individuals, as well as medical and public health teams, will respond to a threat. Again, there are limited empirical data to inform models regarding medical and public response capacity and human behavior in the setting of bioterrorism. As pointed out in Ferguson (2007), there are fundamental limitations in how models can capture the key social parameters of human behavior. The manner in which people alter their behavior in an attempt to reduce their risk when faced with lethal or novel pathogens is difficult to predict and may significantly alter the consequences of an attack. Models are unlikely to capture behaviors that significantly reduce social contact, as seen in Hong Kong and Singapore during the epidemic of severe acute respiratory syndrome (SARS) in 2003, or potential unintentional contact-*increasing* behavior after the occurrence of a widely publicized bioterrorism attack.

Soon after the terrorist attacks of September 11, 2001 (9/11), several prominent infectious-disease modelers undertook studies to assess the likely magnitude of smallpox epidemics under various response strategies. The U.S. government was particularly interested in determining whether, in the aftermath of an attack, vaccination of likely contacts of infected persons ("ring vaccination" or "traced vaccination") would be as effective in containing an outbreak as would mass vaccination. At that time, soon after 9/11, vaccine was in limited supply. The former strategy would require fewer vaccinations and, due to the capability of smallpox vaccine to induce a reaction, would be associated with less morbidity. Despite available quantitative data from past smallpox epidemics, there was considerable disagreement about the likely adequacy of the various responses. It took several years and considerable debate to understand that the differences in models' conclusions rested mainly on assumptions about the timing of transmission relative to symptoms and about the likely speed of the public health response (i.e., the capacity of public health workers to enact targeted versus mass vaccination campaigns) (Cooper, 2006). Substantial differences between public health capabilities in different jurisdictions present more variability. Thus, the site of the attack may significantly influence its consequences. In summary, response logistics matter just as much as epidemiology in determining the outcome of a bioterrorism attack (Kaplan et al., 2003).

Although consequence models are imperfect, they clearly can contribute to planning and mitigation. An appropriate estimate of the damage that the United States could experience is critical to allocating resources and developing mitigation strategies to the numerous possible different threats. "Intuitive judgment" alone is inadequate, as it focuses on only a handful of salient cues (and not necessarily the right ones), often weighed in a simple linear fashion (Hammond, 2006). However, models should take into account the intuitive judgments of informed and experienced health professionals. A "structured discussion" approach is also useful in driving consensus. However, structured discussion is subject to small-group dynamics and may reflect primarily the biases of the most vocal, argumentative, or influential individuals (Janis, 1989). Both intuitive judgment and structured judgment can be useful adjuncts to the modeling process but, like modeling itself, are not sufficiently robust and free of bias and error to stand alone.[5]

Recommendation: The susceptible, exposed, infected, and recovered (SEIR) model adopted by DHS is more complex than can be supported by existing data or

[4]In May (2004), Lord May writes, "The increasing speed and sophistication and ease of use of computers enables an increasingly large number of life scientists who have no substantial background in mathematics to explore 'mathematical models' and draw conclusions about them. Such activity usually consists of representing sensible and evidence-based assumptions as the starting point for a complicated and usually nonlinear dynamical system, assigning particular parameters (often in an arbitrary way), and then letting this complicated system rip. This represents a revolutionary change in such theoretical studies. Until only a decade or two ago, anyone pursuing this kind of activity had to have a solid grounding in mathematics. And that meant that such studies were done by people who had some idea, at an intuitive level, of how the original assumptions related to the emerging graphical display or other conclusions on their computer. Removing this link means that we arguably are seeing an increasingly large body of work in which sweeping conclusions—'emergent phenomena'—are drawn from the alleged working of a mathematical model, without clear understanding of what is actually going on. I think this can be worrying." (p. 790)

Lord May further substantiates his argument in favor of simpler models over complex ones by citing the example of HIV/AIDS models developed in the mid-1980s to estimate the likely demographic impact in some central African countries: "The main unknown at that time was the probability, ß, that an infected individual would infect a susceptible partner. Available data suggested that ß depended relatively little on the number of sexual acts within a partnership. On this basis, we used a relatively simple model to suggest that the future demographic impact of HIV/AIDS could be severe in some such countries. In contrast, the World Health Organization and the Population Council in New York produced models that were much more complex, including very detailed demographic data, but where HIV transmission probability was treated as if for measles, compounding independently and randomly for each individual sex act. Thus, in effect, their models assumed that, knowing nothing of the infective status of individuals, 1 sex act with each of 10 different sex partners was effectively equivalent to 10 acts with 1; our data-governed, but otherwise much simpler, model saw the former as roughly 10 times more risky. So it was not surprising that the later models, apparently 'more realistic' by virtue of their computational complexity, suggested a less gloomy view than ours. Sadly, but understandably, our predictions have proved more reliable." (p. 793)

[5]An encompassing summary of forecasting using expert judgment can be found in Armstrong (2001).

knowledge. DHS should make its SEIR model as simple as possible consistent with existing knowledge.

The complexity of the consequence models presented by DHS seems too great given the data available. The use of a complex model when adequate data are unavailable is probably detrimental to the quality of conclusions, and their use may be dangerously misleading. The complexity compromises the ability to elicit sensible estimates, uncertainty ranges, and correlations in the uncertainty for all of these parameters obtained from subject-matter experts. Hence the uncertainty of the model will likely be incorrectly estimated. In addition, complex models do not lend themselves well to independent validation and verification by other modelers.

OTHER CONSEQUENCES NEED TO BE MODELED

Bioterrorist attacks create direct impacts, which occur immediately after the event, and indirect impacts, which may be much longer term in nature. More specifically, *direct impacts* are damage and losses that can be directly attributed to the attack, such as injuries, loss of life, and damage to property and infrastructure as well as to natural habitats and fish and wildlife populations. *Indirect or secondary impacts* occur over time and include, for instance, family trauma and social disruption, business interruptions, and shortages of critical human services. From a societal point of view, a large number of cases of a disease, perhaps with a long period of latency, can have far more serious implications than the consequences of a large number of deaths from the same disease.[6]

Any measure of health consequences must, at a minimum, combine considerations of morbidity and mortality. Two such measures of "health utility" commonly used are the disability-adjusted life-year (DALY) and the quality-adjusted life-year (QALY). The more commonly used DALY is computed as $(N \times L) + (D \times DW \times DD)$, where N = number of deaths, L = life expectancy in years, D = number of disabilities, DW = disability weight, and DD = duration of disability in years. The disability weight is defined by a panel of clinicians. Some DALY models also apply age-weights and discount rates. The scaling is such that 1 DALY represents the loss of 1 year of equivalent full health.[7]

Estimates of morbidity should also include psychological effects. Fear is a terrorist's "force multiplier." Fears about biological agents, which are not readily identifiable and are generally misunderstood, heighten the real or perceived threats of terrorism. After the anthrax letter attacks in the United States in October 2001, millions of people were made anxious opening their mail; and there have been numerous reports of persons opening a letter or package, finding a powdery substance (later found to be harmless), and having a psychological and physical reaction that required medical attention (Wessely et al., 2001). Silver et al. (2002) point out that the long-term social and psychological effects of a biological attack may be as damaging as the acute ones, that they may remain high for years, and that they may exacerbate preexisting psychiatric disorders and further heighten the risk of mass sociogenic illness. A distrust of medical experts and government officials, who cannot provide blanket assurances of no lasting harm, may result. The response to bioterrorist events may involve the distribution of medical therapeutics and vaccines, isolation of symptomatic individuals, observation of potentially exposed people by public health officials, and other actions that are guaranteed to generate anxiety in the population. The ongoing risk of exposure, possible evacuation from contaminated areas, and the perceived or real risk of death or permanent health consequences are all contributory. If government and public health officials do not properly manage risk communication, there is potential for civil disruption and further business and economic losses. DALY would be an appropriate measure of psychological distress, as it is already used in mental health.

Bioattacks can also have serious environmental consequences. For example, Gruinard Island in the United Kingdom became contaminated with anthrax spores after testing occurred on the island in 1942; the island was quarantined for almost 50 years. Decontamination was finally accomplished in 1986 when, after removal of topsoil, the 520-acre island was soaked with 280 tons of formaldehyde diluted in 2,000 tons of seawater. In the 2001 anthrax attacks in the United States, four major cleanups were required: at the American Media, Inc. (AMI), building in Boca Raton, Florida; the National Broadcasting Company (NBC) offices in New York City; the U.S. Capitol complex in Washington, D.C.; and at two facilities of the U.S. Postal Service.[8]

Agricultural consequences also need to be considered. Economic activity of U.S. agriculture has been estimated to exceed $1 trillion annually, with exports valued in excess of $50 billion. Protecting U.S. agriculture is critical to the global economy and to the ensuring of an adequate and safe food supply in the United States and other countries. Several assessments of agricultural consequences have shown that livestock and poultry populations are vulnerable to biologic attack. The U.S. Department of Agriculture has identified viruses and bacteria capable of causing widescale morbidity and mortality of livestock and poultry that would result in a cessation of international trade and exports costing the United States billions of dollars.[9]

---

[6]For a more detailed discussion of the importance of incorporating indirect or secondary impacts in evaluating alternative programs, see Heinz Center for Science, Economics, and the Environment (1999).

[7]These and other measures of population health are discussed in NRC (1998).

[8]See news.bbc.co.uk/2/hi/uk_news/scotland/1457035.stm. Accessed January 31, 2008.

[9]See frwebgate.access.gpo.gov/cgi-bin/getdoc.cgi?dbname=2005_register&docid=fr18mr05-20.pdf. Accessed January 31, 2008.

Recommendation: While human mortality and the magnitude and duration of morbidity should remain the primary focus of DHS bioterrorism risk analysis, DHS should incorporate other measures of societal loss, including the magnitude and duration of first- and second-order economic loss and environmental and agricultural effects.

Some direct impacts of bioterrorist attacks are relatively easy to quantify because they are easy to measure in dollars: insured losses to homes, businesses, and industry; bridge and highway repairs; equipment replacement or repairs; crop loss; and so on. The costs of other direct impacts and many indirect impacts are less easy to determine and quantify—for example, psychological distress and family instability.

Cost-benefit analysis (CBA) has been proposed as a way of combining direct and indirect effects of alternative programs. If one undertakes CBA, it is necessary to monetize each of the direct and indirect impacts to provide a common metric for ranking the risks of different bioagents. Monetization means assigning values in dollars. Mortality and morbidity (including psychological distress) could be monetized by setting 1 DALY or 1 QALY equal to the "value of a statistical life-year" or to 1 year of income (typically on the order of $50,000). The value of environmental impacts is measured in terms of willingness to pay by using contingent valuation techniques and has been a source of debate by economists over the years.

The total social cost of a bioterrorist attack can be estimated by combining direct and indirect economic costs with the monetization of mortality, morbidity, and environmental costs. Some critics of CBA are unwilling to attach monetary values to life, environmental impacts, or other non-economic consequences from different events. One then has to use other methods of analysis such as cost-effectiveness analysis or multigoal analysis.[10]

REFERENCES

Armstrong, J.S. 2001. *Principles of Forecasting.* Newell, Mass.: Kluwer Academic Publishers.

Boardman, A., D. Greenberg, A. Vining, and D. Weimer. 2001. *Cost-Benefit Analysis: Concepts and Practice.* Upper Saddle River, N.J.: Prentice Hall.

Cooper, B. 2006. "Poxy Models and Rash Decisions." *Proceedings of the National Academy of Sciences* 103(33):12221-12222.

DHS (Department of Homeland Security). 2006. *Bioterrorism Risk Assessment.* Biological Threat Characterization Center of the National Biodefense Analysis and Countermeasures Center. Fort Detrick, Md.

Ferguson, N. 2007. "Capturing Human Behavior." *Nature* 446(7137):733.

Hammond, K.R. 2006. *Beyond Rationality: The Search for Wisdom in a Troubled Time.* New York: Oxford University Press.

Heinz Center for Science, Economics, and the Environment. 1999. *The Hidden Costs of Coastal Hazards: Implications for Risk Assessment and Mitigation.* Washington, D.C.: Island Press.

Janis, I.L. 1989. *Groupthink: Psychological Studies of Policy Decisions and Fiascoes.* Boston: Houghton Mifflin.

Kaplan, E.H., D.L. Craft, and L.M. Wein. 2003. "Analyzing Bioterror Response Logistics: The Case of Smallpox." *Mathematical Biosciences* 185(1):33-72.

May, R.M. 2004. "Uses and Abuses of Mathematics in Biology." *Science* 303(5659):790-793.

NRC (National Research Council). 1998. *Summarizing Population Health: Directions for the Development and Application of Population Metrics.* Washington D.C.: National Academy Press.

Silver R.C., E.A. Holman, D.N. McIntosh, M. Poulin, and V. Gil-Rivas. 2002. "Nationwide Longitudinal Study of Psychological Responses to September 11." *Journal of the American Medical Association* 288(10):1235-1244.

Wessely S., K.C. Hyams, and R. Bartholomew. 2001. "Psychological Implications of Chemical and Biological Weapons." *BMJ* 323(7318):878-879.

[10]For more details on the concepts and practice of cost-benefit analysis and alternative analyses, see Boardman et al. (2001).

7

Improving the Department of Homeland Security's Biological Threat Risk Assessment and Adding Risk Management

[Public Law 107-188:] An Act [t]o improve the ability of the United States to prevent, prepare for, and respond to bioterrorism and other public health emergencies.

—Public Health Security and Bioterrorism Preparedness and Response Act of 2002

THE USE OF PROBABILISTIC EVENT TREES ALONE IS INSUFFICIENT TO MODEL TERRORISM THREATS

Terrorism, especially relatively high-technology bioterrorism, involves intelligent adversaries whose decisions focus on achieving their objectives by responding to the observed and anticipated actions of the opponents. Additionally, the attacker and defender are both limited by technological and resource constraints which influence the choices that they make when committing attacks and arranging defenses. These two aspects are not properly captured by the probabilistic risk assessment adopted by the Department of Homeland Security (DHS) in its Biological Threat Risk Assessment (BTRA) of 2006. Probabilistic risk assessment has its roots in event-tree risk assessments—used to assess failures of engineered systems, purely random hazards, or acts of nature (e.g., storm damage or nuclear reactor accidents).

The excessive complexity of the BTRA assessment of the probability of terrorist decisions is a significant weakness—especially considering that such complexity is not necessary (see Chapter 3). Below, the committee introduces three models in which terrorist decisions are just that, *decisions*—not prior estimates of probabilities. The models represent different trade-offs and assumptions in addressing the risk management problem, but any of the three approaches would improve the methodology currently used by the BTRA or other simple extensions.

Event trees can help focus attention in cases where uncertainty is high or new defense investment can have maximum impact. Event trees also admit flexible calculation—the event outcomes contain the conditional probabilities obtained from any or all of these sources: expert opinion, mathematical equations, or complex simulations. Event trees model sequential time effects, but in the bioterrorism application assessed here, events may occur in parallel or at unknown times. Since credible data are more available and probabilities are more assessable for some conditional distributions than others, the conditional probability distributions are seldom assessed in the chronological order of the event tree. In the BTRA of 2006, however, probability assessment for each event in the tree was done by requiring a chronological ordering of events, using assumptions about dependence on some of the previous events.

Some events of the BTRA of 2006 represent deliberate decisions made by a terrorist, but such events are modeled as random events. Other events represent defensive choices, but these, too, are modeled as random events. The BTRA of 2006 does not properly model intelligent adversaries. Its probability assessment of terrorist decisions is independent of the potential consequences of the attack. As the attacks of September 11, 2001 (9/11) clearly illustrated, terrorists adapt their means and select targets that have a high probability of attaining the consequences that they hope to achieve.

Consideration of terrorist objectives introduces something entirely new to the BTRA, implying a decision theoretic or game-theoretic perspective (Golany et al., 2007). Both decision theory and game theory (including attacker-defender models using mathematical programming) need to be informed by expertise and judgment. In attacker-defender models and other game-theory applications, a rough symmetry between attacker and defender is assumed; that is, what the defender seeks to minimize, the attacker seeks to maximize. This is supported by evidence that al-Qaeda wants to maximize any damage that the United States would rather minimize (e.g., see the captured "Al Qaeda Training Manual," [FAS, 2007]), so if the key U.S. consequence for risk in the BTRA is expected fatalities, then for al-Qaeda it is the first choice to maximize (but other terrorists may have different priorities). Note that if the terrorist uses some other objective but the defender still favors minimizing fatalities, this improves the results for the defender.

The overly complex consequence models used by the BTRA of 2006 to assess fatalities at terminal events are another weakness (Chapter 6). For example, the susceptible, exposed, infected, and recovered (SEIR) model used to estimate the size of a smallpox epidemic started by a single infected individual accounts for every possible disease-transmission pathway. Because of the large uncertainties throughout the model and the uncertainties in the parameters that describe smallpox transmission, the detail and precision reported by this embellishment are illusory.

SEVERAL METHODS ARE AVAILABLE FOR IMPROVED MODELING OF INTELLIGENT ADVERSARIES

Ultimately, the defending of the United States from terrorist attack boils down to choices of investment to prevent, protect against, respond to, and recover from terrorist attacks. The committee has suggested improvements that, if used to simplify, clarify, streamline, and improve the BTRA, would yield more realism, more accuracy, more transparency, and faster computation; additionally the rankings of bioagents by risk would be more credible than those now produced. The BTRA might then be useful to decision makers for purposes of risk management as well as risk assessment and, most important, for exploring homeland security strategic investment choices.

In an earlier recommendation—see Chapter 3, the subsection entitled "The Approach to Determining the Probabilities of Terrorist Decisions Is Incomplete"—the committee advises DHS to model terrorists as intelligent adversaries. Here the committee reinforces that crucial recommendation and provides alternatives for its accomplishment.

Recommendation: In addition to using event trees, DHS should explore alternative models of terrorists as intelligent adversaries who seek to maximize the achievement of their objectives.

The committee does not underestimate the difficulty in producing a dependable and reliable bioterrorism risk analysis that responds to its 13 recommendations. Three appendixes, D, E, and F, in this report present modeling approaches that can be used with the existing BTRA structure to improve the risk analysis. Table 7.1 evaluates these approaches against the 13 recommendations. None of these approaches alone may be an adequate and complete solution to the problem, and any implementation may present unforeseen difficulties. However, the committee believes that a suitable combination of these approaches, and possibly others, is feasible and will yield a risk analysis that satisfies the demands that this committee sees as necessary.

Red Teaming Can Be Used to Understand Intelligent Adversaries

DHS has experience in exercises. But, for instance, although Top Officials 3 (TOPOFF 3) was the most comprehensive terrorism response exercise ever conducted in the United States,[1] it was an exercise in blue (defender) response to attacks scripted in advance. Red teaming can be used for the enhancement of such exercises and for analysis. Red teaming (i.e., terrorist role playing) is a robust and well-understood analysis technique for assessing adversarial risk in complex, dynamic environments. However, red teaming only reveals vulnerabilities and does not directly support decisions about investment trade-offs among different kinds of defenses.

In red-teaming exercises, people are assigned to play the roles of terrorists. It is essential that the adversary's point of view is pursued when considering adversary actions and reactions. The red team must be immersed in enemy culture, tactics, and beliefs. There may also be an opposing blue team playing the roles of defenders. Each of the adversaries has certain resources, certain information, and certain goals. They play out their scenarios, and results can show how bounded human intelligence, nonstandard thinking, and group dynamics may affect the kinds of attacks that are attempted and the kinds of defenses that are successful. By trying to win the encounter for the adversary, the terrorist (or red) team helps to better elucidate defender responses for each adversary course of action.

In principle, red-teaming exercises can become large and complex, depending on the number of different roles, the degree to which the scenario is unstructured, and the number of independent replications that are completed to assess variability in outcome. Nonetheless, this is a relatively inexpensive way for decision makers to learn what they have overlooked about their opponents. Homeland Security Presidential Directive 10 (The White House, 2004) cites red teaming as a technique for better understanding potential enemy actions, and the committee suggests red teaming to DHS as a useful validation test for scenarios favored by the BTRA. Red teaming is just as applicable in improving risk analyses based on decision trees, optimization, and game theory (Reichart, 1998).

Decision Trees Can Model Bioterrorist Threats

In addition to having event nodes whose random outcomes are determined by a probability distribution, a decision tree has decision nodes, whose outcomes are chosen to maximize (or minimize in the case of the defender) the expected consequence from that node forward. The BTRA event tree could be converted to a "bioterrorist decision tree" with four important changes:

- Convert each node representing a terrorist decision into an expected-damage maximizing decision node,
- Assess probabilities of outcomes of random events, rather than probability distributions of outcomes,

[1]Information on TOPOFF exercises is available at www.dhs.gov/xprepresp/programs/editorial_0896.shtm. Accessed September 19, 2007.

- Eliminate nodes representing frequency of attack and potential for multiple attacks, and
- Employ a simple, random-consequence model at each event node in the last stage of the tree.

Called the Bioterrorist Decision Model (BDM), this approach to modeling the scenario presented in the BTRA is developed in Appendix D and briefly described here.

Appendix D presents two figures, Figure D.1 showing the modeling choices made by DHS and Figure D.2 showing alternatives that could be used by the BDM. Using these alternate choices, the Bioterrorist Decision Model can be relatively quickly implemented for bioterrorism risk assessment and risk management because it uses existing techniques (Parnell, 2008), it is a direct modification of the 2006 BTRA event tree, and it uses commercially available, off-the-shelf software. Much of the work done by DHS on segmenting the bioterrorism attack for modeling and on probability assessment and consequence modeling for the BTRA of 2008 can be retained.

The framework represented by the BDM has the potential to resolve all of the major deficiencies that have been identified in the current BTRA. This is a model from the terrorist's point of view. Because U.S. actions and random events are uncertain to the terrorist, these are modeled as events in the decision tree, but terrorist decisions are modeled as decision nodes. Huge BTRA data demands are mitigated by deleting the two most problematic stages (frequency of attack and multiple attacks) and by using probabilities rather than probability distributions for each outcome of each event. The model improves transparency by using commercially available software with extensive graphic visualization and with built-in features to perform sensitivity analyses. Finally, the model can be modified for use in risk management. After risk management decisions are implemented and the probabilities of the random events are changed conditional on these decisions, BDM can be rerun for recalibration.

Attacker-Defender Optimization Can Unify Risk Management, Risk Assessment, and Resource Allocation

Terrorists cannot afford to invest in developing attacks using every major pathogen. Nor can the United States afford every possible defense. Decision makers on both sides have limited resources and seek to optimize their "payoff" subject to these constraints. Appendix E offers an optimization model that unifies risk management, risk assessment, and resource allocation in what is called a "tri-level, defender-attacker-defender" optimization. After 9/11, U.S. law was changed to allow the U.S. Department of Defense to devote resources to defending the United States within its borders, and the authors of Appendix E[2] were asked to convert military "attacker-defender" models in which the United States is the attacker, to "defender-attacker" models in which the United States defends its critical infrastructure from attacks. They have developed more than a hundred such prototypical applications since then, presenting a new one in Appendix E crafted to the exact needs of DHS for bioterrorism.

The three decision stages are these:

1. DHS commits strategic defense investments, chosen from alternate program portfolios each consisting of a compatible set of defense options, to minimize the maximum expected damage from any attack; these investments are of such magnitude that they are necessarily visible to the attacker;
2. The attacker, after observing these defense investments, chooses attack alternative(s) to maximize expected damage; and
3. The defender mitigates damage from the attack(s) with resources already in place as a result of prior strategic investments.

Here, the term damage (to the defender) is used in lieu of, for example, fatalities or other particular consequence.

Using the hypothetical scenario from Chapter 1, one defense option might be to procure 100 million doses of anthrax protective antigen (PA) vaccine, and another to purchase the same number of doses of Russian (STI) live vaccine (see Chapter 5). No defense strategy would include both of these defense options. One attack alternative would be the anthrax attack hypothesized in Chapter 1. Mitigation efforts after this attack would include distributing and using a vaccine, but only if such vaccine has already been put in place by a defense strategy.

This is a very conservative model for the defender because the defender must protect against the worst possible set of attacks. But that is what good management does.

Denote the defense strategy d, the attack alternative a, and the mitigation effort m. A key input is $damage_{d,a}$, the expected damage if defense strategy d has been followed and terrorist attack alternative a is chosen. This is a BTRA output from its suite of consequence models. Denote another input as $mitigate_{d,a,m}$, and suppose that if defense strategy d has been followed and terrorist attack alternative a has been chosen, then mitigation effort m (enabled by d) is put in full force, and the expected damage is reduced by this amount.

Constraints on capital budget for defensive options in any affordable defense strategy govern defender decisions, as do any synergistic or antagonistic interactions among defense options in any defense strategy portfolio that together dictate what $damage_{d,a}$ results, and any other technological or resource limit on the defender. Similarly, limits on terrorist capital and technology are incorporated directly into the attacker model as conventional optimization constraints. *These data are precisely the same as those that the BTRA now presents to subject-matter experts to elicit their opinions*

[2]Gerald G. Brown, W. Matthew Carlyle, and R. Kevin Wood, Department of Operations Research, Naval Postgraduate School, Monterey, California.

TABLE 7.1 Evaluation of Risk Analysis Techniques

Committee Recommendation	Biological Threat Risk Assessment (BTRA) of 2006[a]	Possibly Revised BTRA of 2006[a]	Bioterrorist Decision Tree (Appendix D)	Optimization Models (Appendix E)	Game Theory (Appendix F)[a]
The Department of Homeland Security (DHS) should use an explicit risk analysis lexicon for defining each technical term appearing in its reports and presentations.	*Does not.*	Could be used.	Would be used.		
To assess the probabilities of terrorist decisions, DHS should use elicitation techniques and decision-oriented models that explicitly recognize terrorists as intelligent adversaries who observe U.S. defensive preparations and seek to maximize the achievement of their own objectives.	*Does not.*	*Would require new techniques to replace sole reliance on event trees.*	Terrorist decision nodes replace event nodes, and decision tree is solved to maximize consequences. Consequences can be solved individually or combined using standard decision analysis techniques.	Probabilities of terrorist actions are outputs of optimization model.	Probabilities of terrorist actions are outputs of game theory models.
The event-tree probability elicitation should be simplified by assessing probabilities instead of probability distributions for the outcomes of each event.	*Does not.*	Could be greatly simplified.	Would be done. Probability elicitation is used for events in decision tree.	Would be done. Tree methods are used to calculate expected consequences.	Would be done. Tree methods are used in risk estimates for cost table.
Normalization of BTRA risk assessment results obscures information that is essential for risk-informed decision making. BTRA results should not be normalized.	*Normalizes risk assessment.*	Normalization could be removed.	Not used. Risk assessment would be provided without normalization using cumulative consequence distribution(s).	Not used.	Not used.
Two significant simplifications should be made to the BTRA of 2006 event tree: • DHS should eliminate Stage 1, Frequency of Initiation [of an attack] by Terrorist Group, and Stage 16, Potential for Multiple Attacks; and • DHS should seek opportunities to aggregate some stages of the tree to only those essential to calculate probabilities and consequences with realistic fidelity.	*Does not.*	Stages 1 and 16 could be deleted resulting in a simplified model.	Would be done. Stages 1 and 16 would not be included. Opportunities for aggregated stages would be pursued.	Stages included are optional. Aggregation of stages is mathematically automated.	Would be done. Tree methods are used in risk estimates for cost table.
Subsequent revision of the BTRA should increase emphasis on risk management. An increased focus on risk management will allow the BTRA to better support the risk-informed decisions that homeland security stakeholders are required to make.	*Does not.*	*Would be extremely difficult owing to model complexity.*	Decision trees are routinely used for making resource allocation decisions. Probabilities and consequences would be changed by risk management options.	Primary focus is finding investment portfolio that minimizes expected risk, given that terrorists see these investments before choosing an attack.	*This approach currently lacks a portfolio analysis, which is essential for risk management. But it seems likely that this capability could be added, as duopoly problems.*
DHS should maintain a high level of transparency in risk assessment models, including a comprehensive, clear mathematical document and a complete description of the sources of all input data. The documentation should be sufficient for scientific peer review.	*Does not.*	Could be improved.	Built in with normal decision tree tools, including sensitivity analysis. Bayes nets could increase transparency.	Complete mathematical specification is presented with a complete numerical example.	Complete mathematical specification is presented.

TABLE 7.1 Continued

Committee Recommendation	Biological Threat Risk Assessment (BTRA) of 2006[a]	Possibly Revised BTRA of 2006[a]	Bioterrorist Decision Tree (Appendix D)	Optimization Models (Appendix E)	Game Theory (Appendix F)[a]
Subsequent revision of the BTRA should enable a decision support system that can be run quickly to test the implications of new assumptions and new data and provide insights to decision makers and stakeholders to support risk-informed decision making.	*Does not.*	*Would be extremely difficult owing to model complexity.*	The removal of unnecessary complexity should allow reasonable run times using complete enumeration or Monte Carlo simulation. Insights are provided with normal decision tree analysis tools.	Responsiveness depends on required level of detail. Insights are provided with mathematical programming techniques.	The computing time is not yet known for this kind of approach, operating on realistically large problems.
The BTRA should be broad enough to encompass a variety of bioterrorism threats while allowing for changing situations and new information. DHS should develop a strategy for the rapid assessment of newly recognized and poorly characterized threats.	*Does not.*	Could be done as illustrated in Chapter 5.	Could be done as illustrated in Chapter 5.	Could be done as illustrated in Chapter 5.	Could be done as illustrated in Chapter 5.
The susceptible, exposed, infected, and recovered (SEIR) model adopted by DHS is more complex than can be supported by existing data or knowledge. DHS should make its SEIR model as simple as possible consistent with existing knowledge.	*Does not.*	Could be done.	Would be done.	Would be done.	Would be done.
While human mortality and the magnitude and duration of morbidity should remain the primary focus of DHS bioterrorism risk analysis, DHS should incorporate other measures of societal loss, including the magnitude and duration of first- and second-order economic loss and environmental and agricultural effects.	*Does not.*	Could be done.	Could be done.	Could be done.	Could be done.
In addition to using event trees, DHS should explore alternative models of terrorists as intelligent adversaries who seek to maximize the achievement of their objectives.	*Does not.*	*Would require new techniques to replace sole reliance on event trees.*	Explicitly designed to consider intelligent adversaries.	Explicitly designed to consider intelligent adversaries.	Explicitly designed to consider intelligent adversaries.
The BTRA should not be used as a basis for decision making until the deficiencies noted in this report have been addressed and corrected. DHS should engage an independent, senior technical advisory panel to oversee this task. In its current form, the BTRA should not be used to assess the risk of biological, chemical, or radioactive threats.	*Deficiencies are uncorrected.*	*Analyses for biological, chemical, or radioactive threats would require new techniques for intelligent adversaries to replace sole reliance on event trees.*	Biological, chemical, and radioactive threats could be done with different decision trees for each type of threat. Results would be compared based on consequence distribution(s).	Similar models have been demonstrated for biological, chemical, and radioactive threats, especially when defensive preparations are visible to attacker.	The approach described applies to generic threats, not just biological terrorism.

NOTE: This table evaluates the BTRA of 2006, a possibly revised BTRA, and the three techniques discussed in Appendixes D, E, and F of this report in terms of their responsiveness to the recommendations in the report.

[a]Text in italics represents great difficulty in satisfying the objective or inability to satisfy the objective.

on event probabilities. Here, exactly one defense strategy is chosen, with its defensive option portfolio, but terrorists are allowed to mount fractional attack alternatives, and mitigation efforts may be allocated fractionally within resource limits put in place by a defense strategy. The result is that probabilities emerge as *outputs* from the optimization, that is, as recommended optimal mixed strategies, rather than posing required, subjective inputs from subject-matter experts.

Appendix E presents a simple illustrative example in detail sufficient for any reader with adequate off-the-shelf modeling and optimization software to repeat the exercise. Appendix E also establishes two key theoretical results that permit the full, 18-stage BTRA model to be solved as a tri-level one. Noting that the first (defense strategy) stage is a linear integer program, because choice of strategy is necessarily binary, but that all subsequent stages feature continuous (i.e., perhaps fractional) decisions, mimicking the BTRA of 2006:

- *Result 1:* Any sequence of contiguous continuous stages of defender decisions, or of attacker decisions, can be collapsed into a single stage; and
- *Result 2:* The order of continuous attacker stages, or continuous defender stages, makes no difference to the optimization, so with no loss in generality all continuous attacker stages from the BTRA can be aggregated into a single, second-stage attacker model, and all continuous defender stages can follow in the third stage.

Beyond this, Appendix E shows how to solve this tri-level optimization model at large scale with conventional methods and off-the-shelf software; that is, there is little need for aggregation or sacrifice of essential fidelity to render a smaller model more amenable to solution.

Further insights arise from these models. For instance, as the nation spends more and more money on better and better defenses, terrorists are forced to optimally spread their efforts among more and more attack alternatives, and the United States responds with increasingly diverse mitigation efforts. This dilution of terrorist effort may bring collateral advantage to the defender and afford more and better opportunities for detection and interdiction. (For example, terrorists, even those committed to suicide attacks, fear capture more than death, so the defenders want to increase the apparent risk of detection, interdiction, and capture.)

These models also lend insight into the utility of secrecy and deception. Although strategic defense investments are assumed to be so large that they cannot effectively be hidden (the committee notes without irony that some current DHS efforts can be profiled quickly on the World Wide Web and in the press, and in more detail via open academic literature), the resulting mitigation capabilities are another thing. If the United States knows how well it can mitigate but the terrorist does not, the United States can use this to its advantage. Some such insights are trivial to observe, while others may take additional analysis with optimization. For instance, suppose that $damage_{d,a}$ (i.e., unmitigated risk) is ordered from worst (largest) to best. That is, an ordinal set of (d,a) pairs is created. If the best (largest) mitigation effort for each (d,a) pair would not change this ordering, then there is little sense in taking extraordinary efforts to secret this. Conversely, substantive mitigation abilities that would change this risk ordering are worth keeping secret. See Appendix E for more suggestions about secrecy and insights on deception.

The optimization introduced by Appendix E bears many resemblances to game theory—in particular, to alternating-play, extensive-form games—and there are deep connections not pursued here. Suffice it to say that the optimization proposed accommodates highly detailed technological constraints and resource limits on the opponents (to the extent that they are known), and the solution method offered is completely new and can actually solve these problems at large scale.

Game Theory Models Can Help with Risk Management

Appendix F describes an analysis that combines game theory and statistical risk analysis in the context of a counterbioterrorism example. It is similar to the approach taken in Appendix E, which uses a linear program to solve the underlying game-theory decision making. The main difference is that the method in Appendix F generates many random payoff matrices for the game-theory problem and estimates the proportion of times that a given decision is optimal, as opposed to solving a single game that uses the expected values of the risk distributions as the entries in the payoff matrix. This has the advantage of not overlooking threats that are nearly equal in terms of expected risk, and it provides managers with a comparative view of different defense options. (Appendix F does not address the resource allocation issue treated in Appendix E, but the optimization developed in Appendix E could be transferred to Appendix F.)

More generally, game theory is useful for analyzing the dynamics between terrorist activity and the reactions of defenders when there are interdependencies and weak links in the system. The key point in this model of *interdependent security* is that the incentive which an agent has to invest in risk reduction measures depends on how that agent expects the other agents to invest in security. The agent may change the incentive to invest, or not to invest, depending on the investment of others in security. Consequently, there can be a perverse equilibrium in which no one invests in protection, even though all would be better off if they had incurred this cost. This situation does not have the structure of a prisoner's dilemma game, although it has some similarities (Heal and Kunreuther, 2006). Appendix H develops a more formal model of interdependencies for a two-person game and illustrates situations in which there can be two equilibria—both individuals invest or neither of them takes protective action.

To illustrate in the context of a real-world event, consider the destruction of Pan Am Flight 103 in 1988. In Malta, terrorists checked a bag containing a bomb on Malta Airlines, which had minimal security procedures. The bag was transferred in Frankfurt, Germany, to a Pan American feeder line and then loaded onto Pan Am Flight 103 in London's Heathrow Airport. The transferred piece of luggage was not inspected at either Frankfurt or London, the assumption in each airport being that it was inspected at the point of origin. The bomb was designed to explode above 28,000 feet, a height normally first attained on this route over the Atlantic Ocean. Thus, failures in a peripheral part of the airline network, Malta, compromised the security of a flight leaving from a core hub, London. Terrorists may follow similar behavior with respect to a bioterrorist attack by finding a weak link in the system that could have severe direct and indirect consequences to a much wider population.

The behavior of terrorists is also affected by what their adversaries will do. More specifically, terrorists may respond to security measures by shifting their attention to more vulnerable targets. Keohane and Zeckhauser (2003), Sandler (2005), and Bier et al. (2007) analyze the relationships between the actions of potential victims and the behavior of terrorists. Symmetrically, rather than investing in additional security measures, firms may prefer to move their operations from large cities to less populated areas to reduce the likelihood of an attack. Of course, terrorists may then choose these less protected regions as targets if there is heightened security in the urban areas. Terrorists also may change the nature of their attacks if there are protective measures in place that would make the probability of success of the original option much lower than another course of action (e.g., switching from hijacking to bombing a plane). The impact of endogenous probabilities on the nature of the game-theoretic equilibrium is discussed more fully in Appendix H and in Heal and Kunreuther (2006).

Risk Management Strategies

The three models considered here all treat adversaries as intelligent adversaries that seek to maximize their objectives. Some of the implications are that distributed networks of protection, across different agencies or airlines or firms, may not lead to solutions that are as good as can be obtained with leadership and central direction.

For example, if different defender agents are reluctant to adopt protective measures to reduce the chances of losses from terrorism due to the possibility of contamination from weak links in the system, there may be a role for the private and public sectors to play in addressing this problem. A trade association can play a coordinating role by stipulating that any member must follow certain rules and regulations, including the adoption of security measures. For example, the National Association of Chemical Distributors has developed a code of responsible distribution, mandated third-party auditing of code compliance, and actually terminated membership for noncompliance. Other chemical-infrastructure industry organizations such as the American Chemistry Council, Synthetic Organic Chemical Manufacturers Association, American Petroleum Institute, and National Petrochemical and Refiners Association can also play key roles in this regard.

There may also be a role for governmental standards and regulations coupled with third-party inspections and insurance to enforce these measures. More specifically, third-party inspections coupled with insurance protection can encourage decentralized units in the supply chain to reduce their risks from accidents and disasters. Such a management-based regulatory strategy shifts the focus of decision making from the regulator to individual units that are now required to do their own planning to meet a set of standards or regulations. The combination of third-party inspections in conjunction with private insurance is a powerful combination of two market mechanisms that can convince many units of the advantages of implementing security measures to make their operations more secure. As a result of these units taking action, the remaining ones can be encouraged to comply with the regulations to avoid being caught and fined. This is a form of tipping behavior noted in Appendix H. In other words, without some type of inspection, low-risk units that have adopted risk-reducing measures cannot credibly distinguish themselves from the high-risk ones.

With the delegation of part of the inspection process to the private sector through insurance companies and certified third-party inspectors, a channel would exist through which the low-risk units could speak for themselves. If a unit chose not to be inspected by certified third parties, it would more likely be perceived as high-risk rather than low-risk. If a unit did get inspected and received a seal of approval that it was protecting itself against catastrophic vulnerabilities, the unit would pay a lower insurance premium than that of a unit not undertaking these actions. In this way, the number of audits needed would be reduced because units that had received seals of approval from private third-party inspectors would already be known.

As observed in the safety arena with the National Transportation Safety Board and the U.S. Chemical Safety and Hazard Investigation Board and in the security arena with the 9/11 Commission, an effective system will also independently and publicly investigate when catastrophic failures occur. Investigations examine the root and contributing causes, including the sufficiency of policies, practices, and oversight in the private and public domains. Such future investigations could possibly incorporate a "testing" of the model, or at a minimum provide data about interdependent security.

THE EXISTING BTRA FRAMEWORK SHOULD NOT BE USED FOR THE RISK ANALYSIS OF BIOLOGICAL, CHEMICAL, OR RADIOACTIVE THREATS

National decision makers and DHS leaders will need to allocate scarce resources to prevent, prepare for, and respond to all types of terrorist attacks. Clearly there is a wide variety of potential terrorist attack alternatives (conventional, biological, chemical, and radioactive[3]). Each of these attack alternatives has different attack signatures, detection technologies, and mitigation options. While biological agents can, perhaps, be usefully compared (e.g., by considering whether to invest in vaccines for some specific agent rather than others), there is no analogous comparison for nonbiological agents. For nonbiological agents, the defense of particular locations or facilities against attack and the preparation of mitigation resources should such an attack occur assume a more important role than in the case of biological attack, in which the biological agent used is a primary consideration. In principle, the committee believes that the most simple, meaningful, and useful way to compare biological agents (e.g., anthrax) to chemical agents (e.g., chlorine) and radioactive threats (e.g., a dirty bomb) is by comparison of the potential consequences given a terrorist attack and, when possible, the likelihood of an attack.

However, throughout this report the committee has noted many weaknesses in risk analysis, modeling of intelligent agents, consequence assessment, and presentation of assessment results that it believes make the BTRA of 2006 problematic even for assessing biological agents, let alone other classes of threats. Because of these weaknesses, the rankings produced by the BTRA of 2006 are likely to be biased or skewed by a magnitude that cannot be assessed. Conventional peer review, or periodic reviews by an independent, senior technical advisory panel would almost surely have revealed these BTRA problems earlier. The committee believes that outside oversight will be crucial to correcting these deficiencies.

Recommendation: The BTRA should not be used as a basis for decision making until the deficiencies noted in this report have been addressed and corrected. DHS should engage an independent, senior technical advisory panel to oversee this task. In its current form, the BTRA should not be used to assess the risk of biological, chemical, or radioactive threats.

[3]The committee uses the term "radioactive" to include both "radiological" (i.e., involving radioactive decay such as in a dirty bomb) and "nuclear" (i.e., involving complete fission as in an atomic bomb). Although these two threats are not identical, the committee believes that its recommendations and suggestions *concerning the BTRA methodology used to evaluate the risk of these threats* apply to either.

INTELLIGENT-ADVERSARY RISK ANALYSIS TECHNIQUES CAN BE USED ON RADIOACTIVE AND CHEMICAL THREATS AS WELL AS ON BIOLOGICAL THREATS

Although the committee has recommended that in its present form the BTRA of 2006 and 2008 not be extended to radioactive and chemical risk, it believes that the intelligent-adversary modeling improvements recommended in this report can be applied. Risk management strategies to protect the U.S. chemical infrastructure are discussed in detail in the National Research Council report *Terrorism and the Chemical Infrastructure: Protecting People and Reducing Vulnerabilities* (NRC, 2006). Models for anticipating the actions of intelligent adversaries and for optimizing the allocation of defensive resources can be extended across these areas because all involve similar problems of warning, response, and recovery, and the consequences can be measured in the same consequence units, for example, fatalities. The models suggested here can be applied using risk assessment methods developed specifically for radioactive and chemical risks. Probabilities and consequences in the hypothetical biological scenario used in this report with the probabilities and consequences in radioactive and chemical scenarios can then be compared.

These models can then be used to assess the risk reduction (reduction in probability and/or reduction in consequences) for the resources required for risk management options. Risk management options can then be compared by comparing probability and consequence reduction in each of the three threat areas—biological, chemical, and radioactive. Many risk management alternatives (e.g., vaccines for bioagents, radiation sensors for nuclear threats, and chemical sensors for chemical threats) will only affect the primary threat area. In some cases—for example, recovery options and communication systems—risk management options may result in consequence reductions in all threat areas. In other cases, risk management options may only result in the adversary's shifting or modifying the attack to achieve the same or similar consequences.

Achieving this integrated risk assessment and risk management capability is critical in order for risk-informed decisions to achieve this nation's national security objectives of reducing the threat of weapons of mass destruction.

REFERENCES

Bier, V., S. Oliveros, and L. Samuelson. 2007. "Choosing What to Protect: Strategic Defense Allocation Against an Unknown Attacker." *Journal of Public Economic Theory* 9(4):563-587.

FAS (Federation of American Scientists). 2007. "Al Qaeda Training Manual." Available at www.fas.org/irp/world/para/aqmanual.pdf. Accessed August 23, 2007.

Golany, B., E.H. Kaplan, A. Marmur, and U.G. Rothblum. 2007. "Nature Plays with Dice—Terrorists Do Not: Allocating Resources to Counter Strategic Versus Probabilistic Risks." *European Journal of Operational Research.* In press.

Heal, G., and H. Kunreuther. 2006. "You Can Only Die Once: Interdependent Security in an Uncertain World." In *The Economic Impacts of Terrorist Attacks*, H.W. Richardson, P. Gordon, and J.E. Moore III (eds.). Northampton, Mass.: Edward Elgar.

Keohane, N., and R. Zeckhauser. 2003. "The Ecology of Terror Defense." *Journal of Risk and Uncertainty* 26(2-3):201-229.

NRC (National Research Council). 2006. *Terrorism and the Chemical Infrastructure: Protecting People and Reducing Vulnerabilities*. Washington, D.C.: The National Academies Press.

Parnell, G.S. 2008. "Multi-objective Decision Analysis." *Wiley Handbook of Science and Technology for Homeland Security*. John G. Voeller (ed.). Forthcoming.

Reichart, J.F. 1998. "Adversarial Use of Chemical and Biological Weapons." *Joint Forces Quarterly* 18(Spring):130-133. Available at www.fax.org/irp/threat/cbw/2218.pdf. Accessed October 23, 2007.

Sandler, T. 2005. "Collective Action and Transnational Terrorism." *The World Economy* 26 (6):779-802.

The White House. 2004. Homeland Security Presidential Directive 10 [HSPD-10]: *Biodefense for the 21st Century*. Available at www.fas.org/irp/offdocs/nspd/hspd-10.html. Accessed January 16, 2008.

Appendixes

Appendix A

Lexicon

INTRODUCTION

The lexicon in this appendix, prepared by the Committee on Methodological Improvements to the Department of Homeland Security's Biological Agent Risk Analysis, is intended to be an exemplar of what might be used in any public presentation and discussion of a probabilistic risk analysis and presented as a supplement to this report. Without a clear and consistent use of language in this technical arena, there will be a tendency for conclusions to be misinterpreted and for policy recommendations based on these conclusions to be misguided.

Because many of the terms in this lexicon (Table A.1) are found in everyday usage, often with implications or meanings different from those presented here, it was suggested that the committee also include "lay definitions" in order to provide a comparison and to help in interpreting various loosely written documents and statements made available to the committee (and the public). However, the committee has intentionally *not* done this, in order to avoid giving credence to analyses that might be flawed by improper use or interpretation of various technical terms. The committee recommends that any governmental agency issuing a report on or engaging in a discussion of risk analysis consider using terms as defined in this lexicon, or establish from the beginning reasons for using alternative definitions.

ANALYSIS AND ASSESSMENT

There is an unfortunate (but readily dealt with) inconsistency in usage between two communities importantly involved in understanding the risk of terrorist events: intelligence analysts and risk analysts.

- In the intelligence community it is customary first to gather information about an opponent's intentions and capabilities and then to use this information to present a statement of the current situation. The first step is usually called "analysis" (indeed, employees assigned to information gathering are called intelligence analysts), and the second step is called an "assessment" of the situation.
- The risk and decision community reverses these definitions: the first step of gathering information (in particular, obtaining information about the uncertainty of events and their possible consequences) is usually called "assessment," while the second step—the process of using this information and combining it in such a way that a decision maker can make better decisions—is usually called "analysis."

For this reason, in the lexicon the committee has taken pains to break out the various components of "risk analysis" as used in its report.

ALTERNATIVE DEFINITIONS FOR "RELATIVE RISK"

The term "relative risk" has a well-accepted definition in the biomedical community: "The risk of harm among a population exposed to a potentially damaging substance, compared to the risk amongst an unexposed population."[1] The term may also be used to describe the ratio: {cumulative incidence rate in the exposed population}/{cumulative incidence rate in the unexposed population}. However, the Department of Homeland Security (DHS) has chosen to use the term for a completely different concept. In particular, "relative risk" *for a particular agent* is determined as follows:[2]

- For each agent i an expected consequence $E(C_i)$ is calculated (by Monte Carlo simulation),

[1]R.M. Anderson and R.M. May. 1991. *Infectious Diseases of Humans.* Oxford, United Kingdom: Oxford University Press.

[2]Department of Homeland Security. 2006. *Bioterrorism Risk Assessment.* Biological Threat Characterization Center of the National Biodefense Analysis and Countermeasures Center, Fort Detrick, Md., p. C-95.

- Probability p_i is assigned to the event {agent i will be used},
- An overall "total" expected consequence (or "risk") is computed $R = \sum p_i E(C_i)$,
- The relative risk for agent i is $R_i = p_i E(C_i)/R$.

That is, "relative risk" is the proportion of the total expected risk contributed by a particular agent. Since this definition is quite different from that used by the biomedical community, it presents a major source of potential confusion and misinterpretation, particularly among readers who are knowledgeable in epidemiology.

COMMENTS ON THE CONSTRUCTION AND USE OF THE LEXICON

- Since the committee's primary objective is to provide consistency among the various terms, the terms are cross-referenced as needed. Column 1 provides synonyms and cross-references for the terms defined. It also gives quoted definitions from the DHS document entitled "A Lexicon of Risk Terminology and Methodological Description of the DHS Bioterrorism Risk Assessment" (DHS, 2007).
- References are given in footnotes to the table. Rather than using highly theoretical sources, the committee chose to rely on widely accepted introductory or basic texts[3] or more contemporary but focused references (e.g., Meyer and Booker[4]). Where appropriate, selected Web sites from well-regarded sources have also been used. However, the committee has intentionally avoided the use of glossaries and lexicons readily available on the World Wide Web but developed for promoting commercial software packages, consulting services, and such. These sites are, for the most part, poorly conceived and, more problematic, have not been vetted by any professional independent set of experts, academics, practitioners, or professional societies.
- The main portion of the lexicon (Part A.1.A), although developed for biological risks, can also be appropriately applied to nonbiological (chemical, radioactive, agricultural, and other) threats. The second part of the lexicon (Part A.1.B), specifically, the terms used in susceptible, exposed, infected, and recovered (SEIR) modeling, applies only to biological risk analysis.
- Although the committee recognizes the long philosophical history of the controversy surrounding the nature of uncertainty, it takes the position that, for the purposes of policy development and decision making (the eventual goal of DHS's risk analysis), all uncertainty (subjective, frequency-derived, and so on) must eventually be encoded into probabilities.
- The entry "[None]" in second column, "Committee's Recommended Definition," indicates a conclusion by the committee that it is not necessary (or it is potentially confusing) to provide a definition. Indeed, the committee recommends that such terms not be used in any formal discussion of methods, results, and so on, unless they are used as exemplars of what *not* to say.
- Due to the (committee) process by which the lexicon was developed, it may not include terms that some readers might find important; further, choices among alternative accepted definitions were made where necessary.

REFERENCE

DHS. 2007. "A Lexicon of Risk Terminology and Methodological Description of the DHS Bioterrorism Risk Assessment." Written communication to the Committee on Methodological Improvements to the Department of Homeland Security's Biological Agent Risk Analysis. April 14, 2007.

[3]For example: W. Feller, 1968, *An Introduction to Probability Theory and Its Applications*, New York: Wiley; D.V. Lindley, 1965, *Introduction to Probability and Statistics from a Bayesian Viewpoint; Part 1: Probability*, Cambridge, U.K.: Cambridge University Press; and B. deFinetti, 1974, *Theory of Probability*, Hoboken, N.J.:Wiley.

[4]For example: M.S. Meyer and J.M. Booker, 2001, *Eliciting and Analyzing Expert Judgment: A Practical Guide*, Philadelphia, Pa.: American Statistical Association and the Society for Industrial and Applied Mathematics.

TABLE A.1 Lexicon of Probabilistic Risk Assessment Terms

PART A.1.A TERMS APPLICABLE TO BIOLOGICAL RISKS AND TO OTHER, NONBIOLOGICAL THREATS

Term, with Synonyms, Cross-References, and DHS Lexicon[a] Definitions	Committee's Recommended Definition	Notes, Comments, and References
accuracy See also **precision**.	A measure of agreement between the estimated value of some quantity and its true value. (Adapted from Society for Risk Analysis [SRA] Glossary.[b])	See note under **precision**
agent-conditional expected risk See also **conditional expected risk**.	The **conditional expected risk** computed using probabilities conditional upon the use of a particular agent.	
agent-conditional relative risk See also **agent-conditional expected risk**.	The **conditional relative risk** using probabilities conditional upon the use of a particular agent.	
aleatory probability Synonym: aleatory uncertainty See also **probability**, **epistemic probability**.	"A measure of the uncertainty of an unknown event whose occurrence is governed by some random *physical* phenomena that are either (1) predictable, in principle, with sufficient information (e.g., tossing a die) or (2) essentially unpredictable (radioactive decay)."[c]	
approximation See also **estimation**.	"The result of a computation or assessment that may not be exactly correct, but that is adequate for a particular purpose."[d]	
arc (directed arc) Synonym: branch See also **split fraction**.	In an event tree: an outcome from a preceding event to a subsequent event; in a decision tree: either an action or an outcome from a preceding event to a subsequent event.	
arithmetic average Synonyms: arithmetic mean, sample mean See also **mean**.	The sum of n numbers divided by n.[e,f,g]	The average is a simple arithmetic operation, requiring a set of n numbers. It is often confused with the mean (or expected value), which is a property of a probability distribution. One reason for this confusion is that the average of a set of realizations of a random variable is often a good estimator of the mean of the random variable's distribution.
conditional expected risk See also **agent-conditional expected risk**.	**Expected risk** computed using conditional probabilities.	The conditioning event is typically the choice of agent; however, it could be other events such as good weather, successful manufacture, or ineffective countermeasures.
conditional probability See also **probability**.	The probability of an event supposing (i.e., "conditioned upon") the occurrence of other specified events. In the aleatory theory of probability, the conditional probability of event *A* given event *B* is equal to the probability of the joint occurrence of events *A* and *B* divided by the probability of event *B*, *if the probability of event B is not zero*. (After Feller [1968],[g] DeFinetti [1974],[h] and Lindley [1965].[i])	It is important to note that subjectively assessed probabilities are based on the state of knowledge that holds at the time of the probability assessment.
conditional relative risk	The proportion of the total expected risk contributed by a particular conditioning event.	If $p_i = P\{\text{conditioning event } i\}$ and C_i = expected consequence associated with event i, then total expected risk is $R = \Sigma\, p_i C_i$ and the total conditional relative risk associated with event i is $p_i C_i / R$.

continued

TABLE A.1 Continued

Term, with Synonyms, Cross-References, and DHS Lexicon[a] Definitions	Committee's Recommended Definition	Notes, Comments, and References
conditional risk See also **risk**, **conditional probability**.	Risk computed using conditional probabilities (follows from the definition of conditional probability).	The expected risk associated with a particular agent, as measured by the expected number of deaths, may be conditioned upon (for example) the direction of the wind.
confidence interval See also **uncertainty range**.	A range of values $[a,b]$ determined from a sample, using a predetermined rule chosen such that, in repeated random samples from the same population, the fraction α of computed ranges will include the true value of an unknown parameter. The values a and b are called confidence limits; α is called the confidence coefficient (commonly chosen to be .95 or .99); and $1 - \alpha$ is called the confidence level. (Adapted from SRA.[b])	Confidence intervals should not be interpreted as implying that the parameter itself has a range of values; it has only one value. The confidence limits a and b, being computed from a sample, are random variables, the values of which (for a particular sample) either do or do not include the true value a of the parameter. However, in *repeated* samples, a certain fraction of these intervals will include the parameter, provided that the actual population satisfies the initial hypothesis.
consequence Synonym: outcome	A description of a scenario, in terms of *measurable* factors, that decision makers may consider in assessing preferences over different scenarios; these factors are often random variables. (Adapted from McCormick [1981],[j] with "damage" replaced by consequences.")	For DHS risk analyses, typical and important consequence measures are lives lost, morbidities, direct and indirect dollar losses, and others.
continuous random variable See also **cumulative distribution function**, **probability density function**.	A random variable that has an absolutely continuous cumulative distribution function.[i,e]	
cost-benefit analysis	"A formal quantitative procedure comparing costs and benefits of a proposed act or policy."[b]	SRA also includes in its definition: "To determine a rank ordering of projects to maximize rate of return when available funds are unlimited, the quotient of benefits divided by costs is the appropriate form; to maximize absolute return given limited resources, benefits minus costs is the appropriate form." This method of rank-ordering is inappropriate for risk analysis in that it implies specific (and presumably known) trade-offs between noncommensurable benefits and costs. A better procedure is to plot the costs and benefits associated with each possible decision and then to present the results to decision makers to assess the trade-offs, which may (or may not) result in the linear or multiplicative functions inherent in the cost-benefit computations.
cumulative distribution function (CDF) Synonyms: cumulative distribution, distribution function See also **probability distribution**, **probability density function**, **probability mass function**.	The function $f(x)$ whose value is the probability that a random variable, X, will be less than or equal to a value x; written as $P\{X \le x\}$.[e,g,k]	The cumulative distribution function always exists for any random variable; it is monotonic and nondecreasing in x, and (being a probability) $0 \le P\{X \le x\} \le 1$. If $P\{X \le x\}$ is absolutely continuous in x, then X is called a "continuous" random variable; if it is discontinuous at a finite or countably infinite number of values of x, and constant otherwise, X is called a "discrete" random variable.

TABLE A.1 Continued

Term, with Synonyms, Cross-References, and DHS Lexicon[a] Definitions	Committee's Recommended Definition	Notes, Comments, and References
decision tree See also **event tree**, **fault-tree analysis**.	A tree with event nodes that are random variables or decision nodes that represent decisions of an active agent. Each branch (path of event and decision nodes leading to a terminal node) may have consequences (e.g., in dollars, lives, utility) associated with its terminal node.	The operations used in a decision tree are elementary: expectation over consequences at event nodes and maximization (or minimization) at decision nodes. Decision trees can be infinite (with no terminal nodes, as in recursive game trees) and/or can have intermediate consequences at nonterminal nodes.
directed arc Synonym: branch See also **split fraction**.	In an event tree: an ordered pair of nodes, representing a preceding event, followed by a subsequent event. It is usual to interpret an arc as the outcome of an event. In a decision tree or game tree: an ordered pair of nodes representing either an action or a preceding event, followed by a subsequent action or event or terminal ("payoff") node.	
discrete random variable See also **cumulative distribution function**, **probability mass function**.	"A random variable that has a non-zero probability for only a finite, or countably infinite, set of values."[c]	A probability mass function is used to represent the set of probabilities for all values of a discrete random variable.
epistemic probability Synonym: epistemic uncertainty See also **aleatory probability**, **uncertainty**. DHS Lexicon: "arising from limited state of knowledge"[a]	"A representation of uncertainty about propositions due to incomplete knowledge. Such propositions may be about either past or future events."[c]	Some examples of epistemic probability are (1) the assigning of a probability to the proposition that a proposed law of physics is true; (2) determination of the probability that a terrorist will use a particular agent, based on evidence presented.
estimation (of parameters in probability models) Also see **approximation**.	"A procedure by which sample data are used to assess the value of an unknown quantity."[f]	Estimation procedures are usually based on statistical analyses that address their efficiency, effectiveness, limiting behaviors, degrees of bias, etc. The most common methods of parameter estimation are maximum likelihood and the method of moments. *Bayesian methods* tend to avoid producing estimates and instead treat parameters as unknown quantities, with associated probability distributions.
event See also **random variable, event space**.	A subset of the sample space.[f,g] In a decision or event tree, a random variable whose values are possible outcomes.	Events are the basic building blocks of a probabilistic risk assessment; they are the entities for which probabilities are assessed and/or computed. Event descriptions must be carefully and unambiguously articulated. The terminal event "100 people die"—without making explicit the time frame within which they die, their geographical distribution, their demographics, etc.—is quite different from "100 people die" within the first 48 hours of the attack, all of whom are within 5 km of the city center, 60% of whom are age 65 and older, and so on. The important thing to consider here is that the granularity of probability risk assessment events should be *only as fine as needed* to capture the consequences of the scenarios that include the events.

continued

TABLE A.1 Continued

Term, with Synonyms, Cross-References, and DHS Lexicon[a] Definitions	Committee's Recommended Definition	Notes, Comments, and References
event space Synonym: sample space See also **event**.	The set of all possible outcomes of an experiment or of some (unknown) phenomenon. (After Feller [1968][g] and Statistical Education Through Problem-Solving [STEP] Consortium.[f])	
event tree Synonyms: probability tree, chance tree See also **tree**, **decision tree**, **fault-tree analysis**. DHS Lexicon: "a logic diagram consisting of both decisions and physical events in which the potential outcomes are represented by a finite, complete, discretized set of outcomes (branches). The events are not necessarily consecutive in time and are, in general, not independent."[a]	A tree formed of a sequence of random variables, called events. The branching point at which a new variable is introduced in the tree is called a node. Each node is followed by the possible random variable realizations, called outcomes, and their probability distributions conditional on outcomes of previous random variables in the tree. The outcomes are represented as arcs leading from one event to the next. The joint probability of the intersection of events that constitute a sequence (or scenario) is found by multiplication. A natural way to construct an event tree is to place events in the chronological order in which they occur, if this order is known.[l]	An event tree is essentially a decision tree with the decisions removed or replaced by nodes representing events that are the result of probabilistic decisions (made either by the decision maker or some other agency). If a node in an event tree represents a decision taken by an adversary, then the (conditional) probabilities of the resulting events must be assessed or computed just as those for any other event nodes. Note that some computations (perhaps based on game-theoretic approaches) might produce event probabilities of 0 or 1, associated with "knowing" with certainty what action the adversary will take. There is no need to disallow infinite or continuous outcomes, as the DHS definition would imply.
expected risk Synonym: expected consequences Although "expected risk" is not in the DHS Lexicon, DHS reports and presentations seem to imply synonymy among the terms "risk" (as related to a specific set of events or scenarios), "expected risk," and "total risk."	A summary measure of risk for an event, scenario, etc., as expressed by the **expected value** of any one of the measurable consequences associated with the risk. (Adapted from McCormick [1981][j] with "damage" replaced by "consequence.")	The committee strongly recommends that, wherever possible, the term "expected risk" be replaced by the specific measure of consequences, such as "expected deaths," "expected loss of income," "expected illnesses." If these measures are combined in some functional way, for example via a utility function, then "expected risk" should be replaced by "expected utility." One difficulty with defining "expected risk" is the historical reality that the discipline of probabilistic risk assessment arose from an understanding of the risks associated with nuclear reactors, chemical plants, and such. In these situations, expected risk is defined to be [expected frequency of occurrence of an event] times [expected consequences of that event].
expected value Synonym: expectation See also **mean**.	The first moment of the probability distribution of a random variable X; often denoted as $E(X)$ and defined as $\sum x_i\, p(x_i)$ if X is a discrete random variable and as $\int x f(x) dx$ if X is a continuous random variable.[g,e]	The arithmetic average of random samples taken from the distribution converges to the mean for all sufficiently large sample sizes, under certain conditions. Ironically, in many cases the expected value of a random variable is a numerical value that the random variable can *never take on*. For example, if a random variable X has $P\{X = 0\} = .5$ and $P\{X = 100\} = .5$, then $E(X) = 50$, even though X can only take on values of 0 or 100. There is also a common confusion between expected value and average, due to the fact that, in the limit, as the sample size becomes very large, the average of a set of observations of a random variable will approach the mean of the random variable's probability distribution. (A curious linguistic note: in French the expectation is called *l'esperance*, which in rough translation means "hoped for." Being simply the result of a mathematical operation, it is neither hoped for nor truly "expected.")

TABLE A.1 Continued

Term, with Synonyms, Cross-References, and DHS Lexicon[a] Definitions	Committee's Recommended Definition	Notes, Comments, and References
fault-tree analysis	"A technique by which events that interact to produce other events can be related using simple logical relationships permitting a methodical building of a structure that represents the system."[b]	
frequency DHS Lexicon: "1. The number of events that would be expected to occur in a time period."[a] "2. A rate (with units, #/time)."[a]	"The fraction of events that satisfy some prespecified criterion; a record of how often each value (or set of values) of the variable in question occurs."[f]	The two DHS definitions confound four different ideas: expected value, rate, fraction of *past* events that satisfy some criterion, fraction of *future* events that satisfy some criterion.
in-degree	The number of arcs resulting in an event. In a tree, the in-degree is one for all events, except the initial event, which has an in-degree of 0.	
initial event Synonym: initial node	The first node in an event tree.	
initiating event Synonym: initial event DHS Lexicon: "An action taken by a terrorist organization to begin the process that may culminate in an act of terrorism."[a]	An event with the potential to initiate a sequence of other events leading to undesirable consequences.	
likelihood See also **likelihood function**, **probability**, **uncertainty**.	The likelihood, $L(A \mid D)$, of an event A, given the data D and a specific model, is often taken to be proportional to $P(D \mid A)$, the constant of proportionality being arbitrary.[m]	In informal usage, "likelihood" is often a qualitative description of probability or frequency. However, equally often these descriptions do not satisfy the axioms of probability. For example, "likelihood" has been used by DHS as a "weight" when informally assessing uncertainties, even though the collection of these weights do not add to 1.
likelihood function See also **likelihood**.	A weighting function interpreted as a function of parameters with the random variable(s) replaced by its (their) observed values.[h,n]	The maximum (with respect to the parameter value) of the likelihood function often produces an estimator of the parameter with desirable properties.
mean See also **expected value**, **arithmetic average**.	The first moment of a **probability distribution**, with the same mathematical definition as expected value. The mean is a parameter that represents the central tendency of the distribution. (After Glossary of Statistics Terms,[e] STEP Consortium,[f] Ross [2000],[o] Devore [2000].[k])	See note under **expected value**.
measurement error	"The unexplainable discrepancy between a measurement and the quality that the measurement instrument is intended to measure."[p]	Measurement error is often decomposed into two components: (1) random variation of measurements on objects of identical quality; (2) a systematic error in measurement (e.g., a measurement device may be out of adjustment).

continued

TABLE A.1 Continued

Term, with Synonyms, Cross-References, and DHS Lexicon[a] Definitions	Committee's Recommended Definition	Notes, Comments, and References
model See also **simulation**.	A representation of some portion of the world in a readily manipulable form. A *mathematical* model is an abstraction that uses mathematical language to describe the behavior of system. (Adapted from Wikipedia.[q])	Mathematical models are used to aid our understanding of some aspects of the real world and to aid in decision making. They are also valuable rhetorical tools for presenting the rationale supporting various decisions, since they arguably allow for transparency and reproduction of results by others. However, models are only as good as their (validated) relationship to the real world and within the context for which they are designed. It is wise to remember the advice of George E.P. Box: "All models are wrong, but some may be useful."
node See also **event**.	A representation of an event or decision in a decision tree. A representation of an event in an event tree.	
node-to-node branch See also **course of action**.	An ordered pair of nodes; a course of action leading from a preceding event to a subsequent one.	
normal distribution Synonym: Gaussian distribution	A symmetric "bell-shaped" probability density function, $(1/(\sigma\sqrt{2\pi}))(-\exp((x-\mu)^2/(2\sigma^2)))$, completely characterized by two parameters: mean μ and standard deviation σ.[g,k,o]	The normal distribution commonly used, since (1) it is (with certain conditions) the limiting distribution of the sum of random variables, (2) it has a certain degree of mathematical tractability, (3) there exist many well-known methods for estimating its parameters, and (4) it represents a reasonable fit to data obtained for a wide variety of situations.
normalized risk See also **conditional relative risk, relative risk**.	The proportion of the total expected risk contributed by a particular agent.	
out-degree	The number of directed arcs leaving a node.	
path Synonym: scenario	A sequence of arcs.	
Poisson distribution	A commonly used probability mass function associated with a random variable X = number of events that occur in a given period of time. The formula is $P\{X = x\} = \mu^x e^{-\mu} / x!$, for $x = 0, 1, \ldots$, where the parameter $\mu = E(X)$ is the mean of the distribution.[k]	The Poisson distribution is often used to reflect "randomness" of events over time—$P\{$time between consecutive occurring events $\geq t\} = e^{-\mu t}$, which does not depend on the time of any previous event.
precision See also **accuracy**.	The implied degree of certainty with which a value is stated, as reflected in the number of significant digits used to express the value—the more digits, the more precision. (Adapted from SRA.[b])	Consider two statements assessing "W = Bill Gates's net worth": A precise but inaccurate assessment is "W is \$123,472.89"; an imprecise but accurate assessment is: "W is more than \$8 billion."
probabilistic risk assessment Synonym: risk assessment	An analytical tool that (1) identifies and delineates logical combinations of basic (not analyzed further) events that, if they occur, could lead to an accident (or other undesired event, called the top event); (2) assesses or approximates the probability of the top event from the probabilities of logical combinations of basic events; and (3) assesses the probable consequences associated with occurrence of the top event.	

TABLE A.1 Continued

Term, with Synonyms, Cross-References, and DHS Lexicon[a] Definitions	Committee's Recommended Definition	Notes, Comments, and References
probability See also **likelihood**, **conditional probability**. DHS Lexicon: "1. A probability assignment is a numerical encoding of the relative state of knowledge (Society for Risk Analysis). 2. The subjectivist viewpoint of probability: the analyst's state of knowledge or degree of belief."[a]	One of a set of numerical values between 0 and 1 assigned to a collection of random events (which are subsets of a sample space) in such a way that the assigned numbers obey two axioms: 1. $0 \leq P\{A\} \leq 1$ for any A, and 2. $P\{A\} + P\{B\} = P\{A \cup B\}$ for two mutually exclusive events A and B.[o]	The definition holds for all quantification of uncertainty: subjective or frequentist.
probability density function (PDF)	The derivative of an absolutely continuous cumulative distribution function.[p] For a scalar random variable X, a function f such that, for any two numbers, a and b, with $a \leq b$, $P\{a \leq X \leq b\} = \int_a^b f(x)dx$.	The PDF is the common way to represent the probability distribution of a continuous random variable, because its shape often displays the central tendency (mean) and variability (standard deviation). From its definition, $P\{a \leq X \leq b\}$ is the integral of the PDF between a and b.
probability distribution	See **cumulative distribution function**.	
probability elicitation Synonym: probability assessment	"A process of gathering, structuring, and coding expert judgment (about uncertain events or quantities)."[r]	There are many approaches for probability elicitation, the most common of which are those used for obtaining a priori subjective probabilities. However, in some sense *all* probabilities, even those that result from statistical analysis of large data sets, are subjective and therefore require elicitation. This is because the conditions under which the data have been collected, and the relevance of these conditions to *future* events for which probabilities are desired, are a matter of expert and subjective judgment. Note that the results of probability elicitations are sometimes called probability "assessments" or "assignments."
probability mass function (PMF) See also **discrete random variable**.	A function that gives the probability that a discrete random variable takes on a particular value.[k,o]	
random error See also **measurement error**.	[None]	This term is meaningful *only* in the context of analyzing the results of a particular experiment and therefore should not be used.
random variable See also **event**, **probability distribution**, **continuous random variable**, **discrete random variable**.	"A real valued function defined on a sample (or event) space."[g]	The random variables of interest to a PRA are those that describe the consequences of a particular event. For example, suppose that the event space consists of only three events: A = "100 deaths, 500 illnesses"; B = "0 deaths, 0 illnesses"; C = "75 deaths, 375 illnesses"; and their respective probabilities are $P\{A\} = .3$, $P\{B\} = .2$, $P\{C\} = .5$. Then, if the random variables are defined to be X = "the number of deaths associated with the event space," and Y = "the number of illnesses associated with the event space," this implies $P\{X = 100\} = P\{A\} = .3$; $P\{X = 12\} = 0$ (there are no events with $X = 0$); $P\{Y = 375\} = .5$; $P\{Y/X = .5\} = P\{A\} + P\{C\} = .8$, etc. A probability distribution, constructed on the range of the random variable, can then be used to assign probabilities to events in the event space.

continued

TABLE A.1 Continued

Term, with Synonyms, Cross-References, and DHS Lexicon[a] Definitions	Committee's Recommended Definition	Notes, Comments, and References
relative risk (in an epidemiological context Synonyms: risk ratio; odds ratio	See health terms in Part A.1.B of this table.	
reload capacity	A measure of the ability to introduce a pathogen into more than one country and/or on more than one occasion.	(Formulated by former Navy Secretary Danzig, according to Marc Lipsitch of the Harvard School of Public Health.)
risk See also **expected risk**. DHS Lexicon: "when used in a general sense: The potential for realization of unwanted, adverse consequences to human life, health, property or the environment" [*American Heritage Dictionary*]; "('technical meaning'): The set of triplets of frequency, scenario and consequences, for all scenarios <f, s, c>; ('as the output of quantitative risk assessment'): First moment of the risk probability density function."[a]	"The potential for unwanted, adverse consequences."[b]	It is important to distinguish between the term **risk**, which involves uncertainties, consequences and conditioning statements, and **expected risk**, which combines these factors using the linear additive expectation operation. It is essential to be absolutely clear when using these two these terms. Unfortunately, even SRA's Glossary[b] intermixes them, since after giving the definition in Column 2, it goes on to say, "estimation of risk is usually based on the *expected value* of the conditional probability of the event occurring times the consequence of the event given that it has occurred"[b]—which is technically incorrect as well as misleading. To make things even more confusing, Appendix C3 ("Risk Integration") of the DHS's 2006 report *Bioterrorism Risk Assessment*[5] defines "risk" as "the probability or frequency of an event multiplied by the consequences of the event," which is both inconsistent and technically meaningless.
risk analysis DHS Lexicon: "A detailed examination including risk assessment, risk evaluation and risk management alternatives, performed to understand the nature of unwanted negative consequences to human life, health, property or the environment; an analytical process to provide information regarding undesirable events; the process of quantification of the probabilities and expected consequences for identified risks (from SRA)."[a]	An overall process that involves risk assessment, risk perception, risk communication, and risk management. The hazards to be analyzed (e.g., physical, chemical, radioactive, and biological agents) may result from natural events (e.g., earthquakes and hurricanes), technological events (e.g., chemical accidents), and human activity (e.g., design and operation of engineered systems or attack by a terrorist). (Adapted from SRA.[b])	
risk assessment See also **risk analysis**. DHS Lexicon: "The process of establishing information regarding acceptable levels of a risk and/or levels of risk for an individual, group, society, or the environment. (From SRA)."[a]	The systematic process of identifying hazards and quantifying their potential adverse consequences (magnitude, spatial scale, duration, and intensity) and associated probabilities, including the uncertainties surrounding these estimates. It may include a description of the cause-and-effect links between hazards, the nature of the interdependencies, vulnerabilities, and consequences. (Adapted and expanded from SRA.[b])	

TABLE A.1 Continued

Term, with Synonyms, Cross-References, and DHS Lexicon[a] Definitions	Committee's Recommended Definition	Notes, Comments, and References
risk communication	The process used by risk analysts, decision makers, policy makers, and intelligent adversaries to provide data, information, and knowledge to change the risk perceptions of individuals and organizations and enable them to assess the risk more accurately than they otherwise might.	
risk curve	A graph describing frequency of events as a function of consequences. Alternatively, a curve describing frequency of events with consequences greater than or equal to some level as a function of that level.	
risk estimation See also **risk analysis**. DHS Lexicon: "The scientific determination of the characteristics of risks, usually in as quantitative a way as possible. These include the magnitude, spatial scale, duration and intensity of associated probabilities as well as adverse consequences and their description of the cause and effect links. (from SRA)"[a]	"The determination of the characteristics of risks such as the magnitude, spatial scale, duration, and intensity of adverse consequences and their associated probabilities of the cause-and-effect links."[b]	Although SRA provides a definition, the committee sees no need to include this term in a formal lexicon, since the term "risk" by itself has many connotations and in any event is a random variable which, by definition, cannot be "estimated." There are also many overlaps with "risk assessment."
risk management See also **risk analysis**. DHS Lexicon: "The process of constructing and evaluating strategies for reducing losses from future hazards and dealing with the recovery process should a disaster occur."[a]	The process of constructing, evaluating, implementing, monitoring, and revising strategies for reducing (or distributing) losses from future hazards and dealing with the recovery process should a hazard occur. Risk management strategies include a combination of options such as providing information (i.e., risk communication), economic incentives (e.g., subsidies, fines), insurance, compensation, regulations, and standards. (Adapted and expanded from SRA.[b])	Taken from the definition in the committee's interim report: "In the case of an individual, private sector or public sector organization, these strategies enable them to transfer, mitigate, or accept their perceived risks. Risk management strategies can be evaluated by undertaking cost-benefit analyses to determine the tradeoff between the reduction of risk and the costs of undertaking such measures. In evaluating a risk management strategy one needs to be concerned with the way resources are allocated (i.e. efficiency considerations) as well as the impact of these measures on different stakeholders (i.e. distribution or equity considerations)."[t]
risk perception See also **risk analysis**. DHS Lexicon: "Beliefs held by individuals or organizations about the risks of a hazard. Risk perception is concerned with the psychological and emotional factors, which have been shown to have an enormous impact on behavior."[a]	Beliefs, attitudes, judgments, and perceptions held by individuals, communities, societies, groups, or organizations about the risks of a hazard. Risk perception is concerned with the psychological and emotional factors. Risk perceptions can be influenced by personal knowledge, experience, and beliefs; they can be affected by changing perceptions of the threat, the vulnerabilities, and/or the consequences; they may be influenced by information about hazards, risk assessments, risk policies, and risk management decisions. (Adapted and expanded from SRA.[b])	

continued

TABLE A.1 Continued

Term, with Synonyms, Cross-References, and DHS Lexicon[a] Definitions	Committee's Recommended Definition	Notes, Comments, and References
scenario DHS Lexicon: "One of a possible combination of approaches leading to the execution of an act of terrorism. An end of an event tree."[a]	A complete enumeration of one path on a tree, from the initial event to the terminal node (if any).	
simulation Synonym: Monte Carlo simulation See also **model**.	"A model constructed so that the input of a large number of random variables drawn from defined probability distributions will generate outputs that are representative of the random behavior of a particular system, phenomenon, consequences, etc., of a series of events."[u]	By its inherent nature, each set of "runs" of a simulation represents the outcomes of a series of experiments. Analysis of simulation output data therefore requires a proper experimental design, followed by the use of statistical techniques to estimate parameters, test hypotheses, etc.
split fraction See also **conditional probability**. DHS Lexicon: "For an event, the relative frequency of a branch."[a]	[None]	Presumably this term has been used by DHS to be synonymous with "conditional probability." However, the DHS definition is not consistent with the DHS definition of "frequency," and "relative frequency" is not defined by DHS.
standard deviation See also **variance**.	"The square root of the variance of a distribution."[o]	
terminal event Synonym: terminal node	An event in an event tree or a decision tree with out-degree 0.	
total expected risk Synonym: total risk	The probability-weighted sum of expected risks associated with all agents. (Implied by DHS usage).	It is preferable that "total risk" should depend on the specific context: the consequence (deaths, utility, etc.) and the events over which the sum is taken (e.g., agents, other conditioning events, etc.). For example, if $p_i = P\{\text{conditioning event } i\}$, and C_i is the expected consequence associated with event *i*, then total expected risk is $R = \Sigma\, p_i C_i$.
tree See also **event tree, decision tree**.	A connected acyclic directed graph with exactly one distinguished (root) node with in-degree 0, and every other node with in-degree 1.	
uncertainty See also **probability**. DHS Lexicon: "Two types of uncertainty are considered and treated differently: aleatory uncertainty—arising from variability (e.g., weather variability); epistemic uncertainty—arising from limited state of knowledge."[a]	The condition of being unsure about something; a lack of assurance or conviction.[d]	The formal definition of "uncertainty" is really not important to the understanding of any PRA method. However, having a clear and agreed on definition of the uses to which any quantification of "uncertainty" is put is crucial. The DHS Lexicon definitions are neither clear nor agreed on, and in fact they confuse the notion of "uncertainty" with the various methods used to *quantify* it in a useful way.

TABLE A.1 Continued

Term, with Synonyms, Cross-References, and DHS Lexicon[a] Definitions	Committee's Recommended Definition	Notes, Comments, and References
uncertainty range DHS Lexicon: "Typically, a confidence interval. For the common definition of risk given above [presumably 'expected consequences per unit time or within a time interval,' but not shown in this table since the committee does not display 'lay definitions'], 'the confidence interval associated with the epistemic uncertainty'."[a]	[None]	Depending on the context, DHS apparently means either (1) a range of probabilities associated with a particular event, scenario, etc.—possibly due to disagreements among subject-matter experts, outputs of a simulation or analytical model, or results of an experiment, etc.; or (2) the range of uncertain outcomes associated with a particular event, scenario, etc.
utility Synonym: utility function DHS Lexicon: "function that transforms measures of consequences into a number."[a]	A real valued function of a consequence.	In economics, "utility" captures "relative happiness" or satisfaction gained by goods and services. In decision analysis, "utility" captures returns to scale and risk preference. In both cases, the assessment of utility values (and hence utility functions) for consequences is an inherently subjective exercise and so depends on the individual (or organization) confronting the possible consequences. Formally, let *A* be the most-preferred possible outcome of a risky prospect, *B* be the least-preferred, and *C* be any other outcome. If a decision maker is indifferent between *C* and a prospect having probability *u* of getting *A* and probability $(1 - u)$ of getting *B*, then *u* is defined as the (von Neumann-Morgenstern) utility of *C*.
variance See also **standard deviation**.	The second moment of a probability distribution, defined as $E(X - \mu)^2$, where μ is the first moment of the random variable *X*.	The variance is a common measure of variability around the mean of a distribution. Its square root, the standard deviation, having dimensional units of the random variable, is a more intuitively meaningful measure of dispersion from the mean.
weight of evidence See also **probability risk assessment**.	The logarithm of $K = P\{x \mid A\} / P\{x \mid B\}$, where *x* is a realization of a random variable, and *A* and *B* are alternative hypotheses. (*K* is also called the likelihood ratio.)[q]	A nonstandard and nonstatistical definition, used by some analysts but not recommended, is as follows: "An elicitation of uncertainty that results in a non-normalized set of numbers which can be normalized (by dividing by the sum over all possible events) to produce probabilities." In some statistical usage, the "weight of evidence" is defined to be 10 times the log-likelihood ratio. More generally, a loosely defined or undefined term indicating the extent to which studies are judged to support a conclusion.

continued

TABLE A.1 Continued

PART A.1.B TERMS USED IN SUSCEPTIBLE, EXPOSED, INFECTED, AND RECOVERED (SEIR) MODELING AND APPLICABLE ONLY TO BIOLOGICAL RISK ANALYSIS

Term, with Synonyms, Cross-References, and DHS Lexicon[a] Definitions	Committee's Recommended Definition	Notes, Comments, and References
Bioshield	A federal program authorized in 2004 to improve medical countermeasures protecting Americans against a chemical, biological, radiological, or nuclear (CBRN) attack.	For more information see http://www.whitehouse.gov/infocus/bioshield/.
contagious DHS Lexicon: "infected and capable of spreading disease."[a]	A person who is infected and capable of transmitting an infectious agent to another host. (Adapted from Thomas and Weber [2001][v].)	This can be used as an adjective or noun, but most often, in the modeling context, as a noun.
dose	The amount (or concentration) of desired matter or energy deposited at the site of effect. (Adapted from SRA.[b])	
exposed See also **infected**. DHS Lexicon: "population who came in contact with the infectious agent or toxin and received an infectious dose."[a]	A person or population that came in contact with the infectious agent or toxin.	For SEIR modeling, but generally not other usage, "exposed" includes only those who received an infectious dose. This can be used as an adjective or noun, but most often, in the modeling context, as a noun.
ill DHS Lexicon: "infected or intoxicated population showing symptoms."[a]	Infected or intoxicated population showing clinical signs and symptoms.	This can be used as an adjective or noun, but most often, in the modeling context, as a noun.
infected DHS Lexicon: "population that has been exposed and received an infectious dose."[a]	An individual or population that has an infectious agent enter and multiply in its tissues.[v]	This can be used as an adjective or noun, but most often, in the modeling context, as a noun.
infectious dose X (IDX)	A dose that is expected to lead to the infection of *X* percent of individuals exposed.	Typically $X = 50$, but it is sometimes set to 10, 90, or other values, depending on the intent of the analysis.
intoxicated DHS Lexicon: "population that has been exposed and received a toxic dose of a toxin."[a]	Population that has been exposed to a threshold amount of toxin and will become ill in the absence of intervention.	
lethal concentration X (LCX)	A concentration that is calculated to kill *X* percent of a population. (Adapted from SRA.[b])	
lethal dose X (LDX)	A dose that is expected to kill *X* percent of a population in the absence of medical intervention(s).[b]	Typically $X = 50$, but it is sometimes set to 10, 90, or other values, depending on the intent of the analysis.
R_0 Synonym: basic reproduction number See also **R**.	The mean number of secondary cases of infection to which one primary case gives rise throughout its infectious period, if introduced into a population consisting solely of susceptible individuals. (Adapted from Anderson and May [1991].[w])	R_0 is a property of the pathogen. R_0 is a theoretical number and does not hold if the population is not entirely susceptible, or even in the case where there is more than 1 contagious person (since the entire population is not susceptible).

TABLE A.1 Continued

Term, with Synonyms, Cross-References, and DHS Lexicon[a] Definitions	Committee's Recommended Definition	Notes, Comments, and References
R Synonym: effective reproduction number See also $\mathbf{R_0}$.	The number of secondary cases of infection to which a single contagious case gives rise throughout its infectious period, in a host population where not all persons are susceptible.	R is a property of both the pathogen and the population's relative susceptibility. Under conditions of stable endemic infection, R = 1. Note that the R value can and does change as the outbreak progresses. The change in R may be due to reduction in the susceptible population, through natural infections, changes in social behavior, or medical interventions.
relative risk Synonyms: risk ratio; odds ratio	(Biomedical context) The ratio of the risk of disease or death among the exposed to the risk among the unexposed.	
removed DHS Lexicon: "population that has recovered or died."[a]	Population that has recovered, has been successfully immunized, or has died.	"Removed" may also include vaccinated individuals in some models.
susceptible DHS Lexicon: "population who [sic] is at risk of becoming infected if exposed to an infectious agent."[a]	Individual or population who, if exposed to an infectious agent, could become infected.	This can be used as an adjective or noun, but most often, in the modeling context, as a noun.

[a]Department of Homeland Security. 2007. "A Lexicon of Risk Terminology and Methodological Description of DHS Bioterrorism Risk Assessment." April 14.
[b]Society for Risk Analysis (SRA), Glossary of Risk Analysis Terms. Available at sra.org/resources_glossary.php. Accessed Feb. 22, 2008.
[c]Cornell LCS Statistics Laboratory. See http://instruct1.cit.cornell.edu:8000/courses/statslab/Stuff/index.php. Accessed Feb. 22, 2008.
[d]*American Heritage Dictionary.* 2000. Boston: Houghton, Mifflin.
[e]Glossary of Statistics Terms. Available at www.stat.berkeley.edu/users/stark/SticiGui/Text/gloss.htm. Accessed Feb. 22, 2008.
[f]Statistical Education Through Problem Solving [STEP] Consortium. Available at www.stats.gla.ac.uk/steps/index.html. Accessed Feb. 22, 2008.
[g]W. Feller. 1968. *An Introduction to Probability Theory and Its Applications.* New York, N.Y.: Wiley.
[h]B. DeFinetti. 1974. *Theory of Probability.* Hoboken, N.J.: Wiley.
[i]D.V. Lindley. 1965. *Introduction to Probability and Statistics from a Bayesian Viewpoint; Part 1: Probability.* Cambridge. U.K.: Cambridge University Press.
[j]N.J. McCormick. 1981. *Reliability and Risk Analysis.* San Diego, Calif.: Academic Press.
[k]J.L. Devore. 2000. *Probability and Statistics for Engineering and the Sciences.* Pacific Grove, Calif.: Duxbury Press.
[l]E. Paté-Cornell. 1983. "Fault Trees vs. Event Trees in Reliability Analysis." *Risk Analysis* 4(3):177-186.
[m]A.F.W. Edwards. 1992. *Likelihood.* Baltimore, Md.: Johns Hopkins University Press.
[n]The White House. 2004. Available at www.whitehouse.gov/infocus/bioshield. Accessed Feb. 22, 2008.
[o]S.M. Ross. 2000. *Introduction to Probability Models.* New York, N.Y.: Academic Press.
[p]Duke University. 1998. *Statistical and Data Analysis for Biological Sciences.* Available at isds.duke.edu/courses/Fall98/sta210b/terms.html. Accessed Feb. 22, 2008.
[q]Wikipedia: "Statistics." Available at en.wikipedia.org/wiki/Statisics. Accessed Feb. 22, 2008.
[r]M.S. Meyer and J.M. Booker. 1991. *Eliciting and Analyzing Expert Judgment.* Los Alamos, N.M.: Los Alamos National Laboratory.
[s]DHS (Department of Homeland Security). 2006. *Bioterrorism Risk Assessment.* Biological Threat Characterization Center of the National Biodefense Analysis and Countermeasures Center. Fort Detrick, Md.
[t]National Research Council. 2006. *Interim Report on Methodological Improvements to the Department of homeland Security's Biological Agent Risk Analysis.* Washington, D.C.: The National Academies Press.
[u]E.J. Henley and H. Kumamoto. 1981. *Reliability Engineering and Risk Assessment.* Upper Saddle River, N.J.: Prentice-Hall.
[v]J.C. Thomas and D.J. Weber. 2001. *Epidemiologic Methods for the Study of Infectious Diseases.* Oxford, U.K.: Oxford University Press.
[w]R.M. Anderson and R.M. May. 1991. *Infectious Diseases of Humans.* Oxford, U.K.: Oxford University Press.

Appendix B

Mathematical Characterization of the Biological Threat Risk Assessment Event Tree and Risk Assessment

Gerald G. Brown, Ph.D.
Distinguished Professor of Operations Research
Naval Postgraduate School, Monterey, California

An event tree can be defined as a directed-out-tree (i.e., a connected di-graph that contains no cycle with exactly one, distinguished, root node with in-degree 0, and every other node with in-degree 1).[1] Each node represents some event, and each directed out-arc represents a randomly-chosen outcome that selects a successor event node. Every directed path in this tree starts with the root node, and ends at a node with out-degree zero (a leaf node). Each directed path from the root node to a leaf node in the event tree represents a possible sequence of alternating events and outcomes (i.e., a scenario).

Figure B.1 defines the Biological Threat Risk Assessment (BTRA) event tree mathematically and shows how to solve for all path probabilities. This event tree is a restriction of a completely general one: This tree consists of successive stages, or echelons of events, with each stage restricted to offer the same branch opportunities.

Figure B.2 defines the BTRA risk analysis mathematically.

If we attach a set of mutually-exclusive, exhaustive probabilities to the arcs branching out of each node, we can trace each directed path in the event tree and reckon its joint probability of selection by multiplying the successive arc selection probabilities on the path. Note that we need not assume independence among successive probabilities, and can in fact condition each arc probability on all prior outcomes in its path.

If we associate a consequence (i.e., a measured outcome) with each end state node, we can assess the total expected consequence of each path by multiplying this consequence by its path probability. We can also generalize to a distribution of consequences for each end state node, and accumulate an expected distribution of consequences.

Many of the scenario paths terminate early (e.g., due to interdiction), so the actual number of paths terminating with non-zero consequences is in the thousands, rather than billions.

The distributions of consequences for all scenarios (paths) share the same "bin structure" (discrete intervals), and random sampling of paths can be used to induce a random sampling of consequence distribution. From this expected consequence distribution, we can estimate, for instance, the 5th and 95th percentiles.

[1]See, for example, R. Ahuja, T. Magnanti, and J. Orlin, 1993, *Network Flows: Theory, Algorithms, and Applications*, Upper Saddle River, N.J.: Prentice Hall, Chapter 2.

Index Use [cardinality]

$g = \{1,2,\ldots,G\}$ — ordinal set of successive stages of events leading from initiation of attack planning to final attack consequence. (alias g') [18]

$a_g \in A_g$ — outcome at stage $g < G$ [2-28]

$p_g = \{a_1,\ldots,a_g\} \in P_g = \{a_1 \times \ldots \times a_g\}$ — sequence of outcomes chosen through stage $g < G$ $\prod_{g'<g} |a_{g'}|$ [10^9]

Given Data [units]

$branch\_pr_{p_g}(a_g)$ — probability that at stage p_g outcome a_g is chosen. This probability may depend on every outcome in path $p_g = \{a_1,\ldots,a_g\}$. [probability]

Computed Parameters [units]

$path\_pr(p_g)$ — probability of path p_g [probability]

Computation

$$path\_pr(p_{g+1}) = branch\_pr_{p_g}(a_g) \times \left[path\_pr(p_g) \right]_{g>1}, \ \forall a_g \in A_g, \ g = \{1,\ldots,G-1\}$$

FIGURE B.1 Mathematical definition of BTRA event tree and solution for tree probabilities. This defines a BTRA event tree and shows how to completely evaluate all probabilities for every path. This definition applies whether or not the tree includes all agents, or just one of them.

Additional Index Use [cardinality]

$c \in A_{G-1} \equiv C$ — set of final consequences, outcomes in penultimate stage $G - 1$ [10]

Additional Data [units]

$cost_c$ — cost of consequence c [cost]

Computed Parameters [units]

$cost\_pr(c)$ — probability of consequence c with $cost_c$ [cost]

R — total risk (i.e., expected cost) [cost]

Computation

$$cost\_pr(c) = \sum_{P_g \in P_{G-1}} path\_pr(p_g) \times cost\_pr(c), \ \forall c \in C$$

$$R = \sum_{c \in C} cost_c \times cost\_pr(c) = \sum_{\substack{c \in C, \\ P_g \in P_{G-1}}} cost_c \times path\_pr(p_g) \times cost\_pr(c)$$

FIGURE B.2 Mathematical definition of BTRA risk analysis. This shows how to completely evaluate all cost consequences and risk (expected cost). The paths here have one extra, final stage that BTRA does not: This stage eliminates the necessity for separate notation for consequence distributions, with each of its outcomes resulting in a scalar cost consequence. A Monte Carlo sampling to estimate these computed parameters would proceed by randomly selecting a path $p_{G-1}=\{a_1,a_2,\cdots,a_{G-1}\}$ (the probability of this path could be computed by $\prod_{g<G} branch\_prob_{p_g}(a_g)$, but this is not essential) and collecting this result as a sample statistic.

Appendix C

Computational Example Illustrating the Replacement of a Joint Distribution of Arc Probabilities with Marginal Expected Values of Individual Arc Probabilities

Alyson Wilson, Ph.D.
Technical Staff Member, Statistical Sciences Group
Los Alamos National Laboratory, Los Alamos, New Mexico

Stephen Pollock, Ph.D.
Professor Emeritus, Department of Industrial and Operations Engineering
University of Michigan, Ann Arbor, Michigan

This appendix illustrates two suggestions from Chapter 3 with illustrative *R* code. In particular, we consider:

- the addition of an 18th stage to represent distributions of alternate consequences; and
- replacing distributions of arc probabilities by expected values of the probabilities.

We work from the event tree in Figure C.1. For simplicity, we assume a single initiating event. For concreteness, we assign uncertainty distributions to each of the arc probabilities:

$P_{A1} \sim$ Beta(2,2);
$P_{T1} \sim$ Beta(4,1); and
$P_{T2} \sim$ Beta(3,2).

In addition, we know the distributional form of each consequence distribution. Using the notation $c(x \mid s_1)$ to denote the consequence distribution associated with the first arc, we assign the following distributions to consequences:

$c(x \mid s_1) \sim$ Gamma(8000,2);
$c(x \mid s_2) \sim$ Gamma(4500,1);
$c(x \mid s_3) \sim$ Gamma(10000,2); and
$c(x \mid s_4) \sim$ Gamma(5500,1).

We would like to know the form of the risk distribution. Summary statistics from this distribution (5th percentile, mean, 95th percentile) are used to summarize risk and present analyses in the Biological Threat Risk Assessment (BTRA) of 2006.

A simple way to simulate from the risk distribution is as follows:

- Repeat *n* times;
- Sample from each arc probability;
- Calculate the probabilities for each scenario;
- Choose a scenario using the calculated probabilities;
- Sample from the consequence distribution for that scenario;
- The *n* samples constitute a sample from the risk distribution; and
- Summarize these samples using a histogram, empirical quantiles, and sample mean.

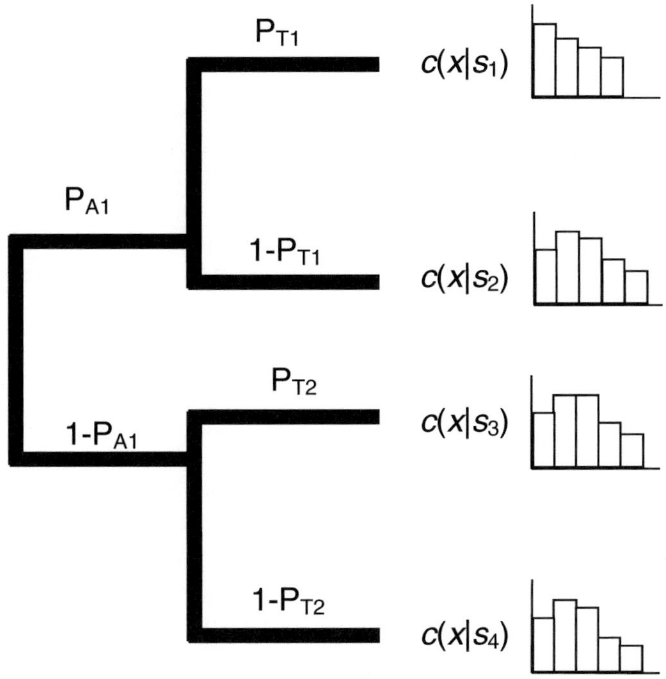

FIGURE C.1 A simple event tree for two successive stages (events), each with two outcomes. For this example, each path through the tree represents a unique scenario with its own consequence distribution.

R code implementing this algorithm follows.

```
n <- 1000000
consq <- rep(0,n)

for (i in 1:n) {
  pa1 <- rbeta(1,2,2)
  pt1 <- rbeta(1,4,1)
  pt2 <- rbeta(1,3,2)

  s1p <- pa1*pt1
  s2p <- pa1*(1-pt1)
  s3p <- (1-pa1)*pt2
  s4p <- (1-pa1)*(1-pt2)

  scen <- rmultinom(1,1,c(s1p,s2p,s3p,s4p))
  if (scen[1] == 1) consq[i] <- rgamma(1,8000,2)
  if (scen[2] == 1) consq[i] <- rgamma(1,4500,1)
  if (scen[3] == 1) consq[i] <- rgamma(1,10000,2)
  if (scen[4] == 1) consq[i] <- rgamma(1,5500,1)
}

hist(consq,freq=F,main="",xlim=c(3500,6000),
  xlab="Consequence Distribution",ylim=c(0,0.0035))
lines(density(consq))
quantile(consq,c(0.05,0.95))
mean(consq)
```

The histogram summarizing the risk distribution from this approach is given in Figure C.2, with an overlay of a kernel density estimator of the risk distribution as the solid line.

The histogram and solid black line result from brute force sampling from the arc probability distributions and the consequence distributions. The line with circles is the estimate from the methodology employed in the BTRA of 2006, which can also produce risk curves. The line with triangles is the estimate from a greatly simplified algorithm that uses only the marginal expected values of individual arc probabilities and simulations from the consequence distributions. The line with crosses is calculated assuming a parametric (or tabular) form is known for the consequence distributions and requires no simulation. Notice the good agreement between the four estimates.

For an event tree as complex as the one presented in the BTRA, this approach is infeasible. As we understand it, the approach implemented in the BTRA is as follows:

- Draw 500 samples from each arc probability;
- Calculate 500 sets of scenario probabilities;
- Draw 1000 samples from each consequence distribution;
- Represent each consequence distribution as a histogram;
- For each of the 500 sets of scenario probabilities, calculate a weighted average of the mass in each bin of the histogram, and call this one "sampled risk curve";
- Calculate the average over all 500 risk curves. Use this as an approximation to the risk distribution and calculate the mean, 5th percentile, and 95th percentile; and
- Also calculate the 5th and 95th percentiles for the entire set of risk curves.

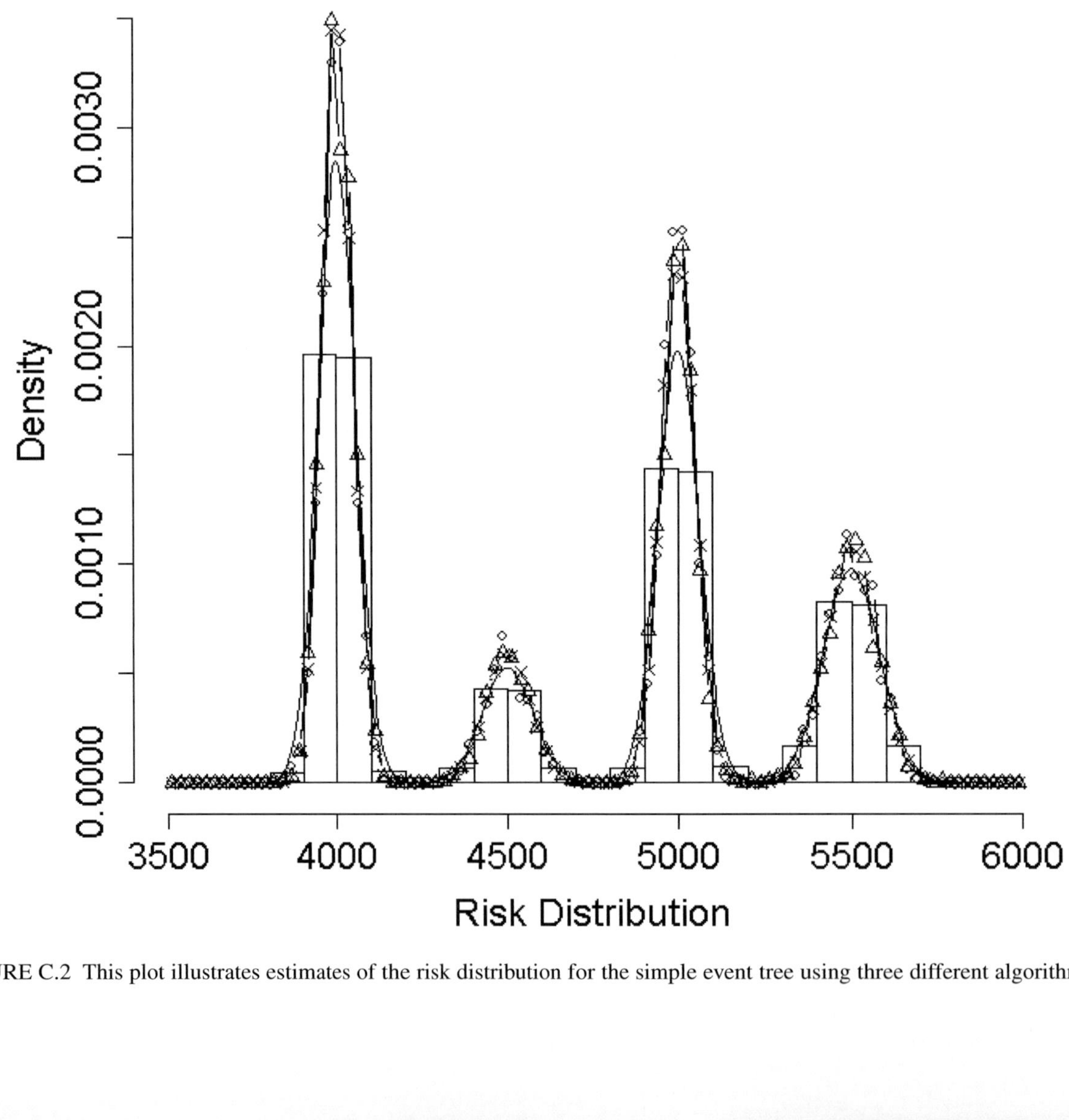

FIGURE C.2 This plot illustrates estimates of the risk distribution for the simple event tree using three different algorithms.

R code implementing this algorithm follows.

```
nsampbr <- 500

pa1 <- rbeta(nsampbr,2,2)
pt1 <- rbeta(nsampbr,4,1)
pt2 <- rbeta(nsampbr,3,2)

s1p <- pa1*pt1
s2p <- pa1*(1-pt1)
s3p <- (1-pa1)*pt2
s4p <- (1-pa1)*(1-pt2)

nsampc <- 1000

cs1 <- rgamma(nsampc,8000,2)
cs2 <- rgamma(nsampc,4500,1)
cs3 <- rgamma(nsampc,10000,2)
cs4 <- rgamma(nsampc,5500,1)

bh1 <- hist(cs1,breaks=seq(3500,6000,length=101),plot=F)$density
bh2 <- hist(cs2,breaks=seq(3500,6000,length=101),plot=F)$density
bh3 <- hist(cs3,breaks=seq(3500,6000,length=101),plot=F)$density
bh4 <- hist(cs4,breaks=seq(3500,6000,length=101),plot=F)$density

qdm <- matrix(0,nsampbr,100)
for (i in 1:nsampbr) {
  qdm[i,] <- s1p[i]*bh1 + s2p[i]*bh2 + s3p[i]*bh3 + s4p[i]*bh4
}

qdmean <- apply(qdm,2,mean)
qd5 <- apply(qdm,2,quantile,c(0.05))
qd95 <- apply(qdm,2,quantile,c(0.95))

x <- seq(3512.5,5987.5,by=25)
points(x,qdmean,type="b",pch=1)
```

The estimated risk distribution from this approach is given as the line with circles in Figure C.2.

As shown in Chapter 3, the risk distribution can be calculated without sampling from the arc probability distributions. For an event tree the size of the one used in the BTRA of 2006, this represents a significant computational simplification. What is lost in the simplification is the family of risk curves—i.e., one curve for each possible outcome. However, no analysis in the BTRA of 2006 and no improvement in analysis recommended by the committee can make meaningful use of the information available in the family of risk curves, beyond that provided by their expectation.

Further, given the improvements proposed for the BTRA to incorporate additional consequence measures and utility functions, the committee does not see upcoming analyses that require the family of risk curves.

Consider the following simplified algorithm:

- Draw 1000 samples from each consequence distribution;
- Represent each consequence distribution as a histogram; and
- Calculate a weighted average of the mass in each bin of the histogram using the expected arc probabilities and use this as the estimated risk distribution.

R code implementing this algorithm follows.

```
ms1p <- (0.5)*(0.8)
ms2p <- (0.5)*(0.2)
ms3p <- (0.5)*(0.6)
ms4p <- (0.5)*(0.4)

nsampc <- 1000

cs1 <- rgamma(nsampc,8000,2)
cs2 <- rgamma(nsampc,4500,1)
cs3 <- rgamma(nsampc,10000,2)
cs4 <- rgamma(nsampc,5500,1)

bh1 <- hist(cs1,breaks=seq(3500,6000,length=101),plot=F)$density
bh2 <- hist(cs2,breaks=seq(3500,6000,length=101),plot=F)$density
bh3 <- hist(cs3,breaks=seq(3500,6000,length=101),plot=F)$density
bh4 <- hist(cs4,breaks=seq(3500,6000,length=101),plot=F)$density

erd <- ms1p*bh1 + ms2p*bh2 + ms3p*bh3 + ms4p*bh4

x <- seq(3512.5,5987.5,by=25)
points(x,erd,type="b",pch=2)
```

The estimated risk distribution from this approach is given as the line with triangles in Figure C.2.

If the conditional consequence distributions are given in parametric form, or in numerical look-up tables, calculation of the risk distribution can be done *exactly,* without resorting to estimating these distributions from the outputs of Monte Carlo simulations. This method is simply:

- Calculate the expected arc probabilities; and
- Calculate the weighted average of the consequence distributions.

```
ms1p <- (0.5)*(0.8)
ms2p <- (0.5)*(0.2)
ms3p <- (0.5)*(0.6)
ms4p <- (0.5)*(0.4)

x <- seq(3512.5,5987.5,by=25)
points(x,ms1p*dgamma(x,8000,2) + ms2p*dgamma(x,4500,1) + ms3p*
dgamma(x,10000,2) + ms4p*dgamma(x,5500,1),type="b",pch=4)
```

The risk distribution (exact, and *not an estimate*) obtained using this approach is given as the line with crosses in Figure C.2. This computation is both trivial and fast.

Appendix D

Bioterrorism Risk Analysis with Decision Trees

Gregory S. Parnell, Ph.D.
Professor, Department of Systems Engineering
United States Military Academy, West Point, New York

INTRODUCTION

The foundational risk analysis method used by the Department of Homeland Security (DHS) Biological Threat Risk Assessment (BTRA) methodology is event trees. Event trees are a proven probabilistic risk analysis technique that has been effectively used for risk analysis of natural and man-made hazards (Dillon-Merrill, Parnell, and Buckshaw, 2007). The body of this report has shown weaknesses in the use of event trees to model terrorist actions since event trees do not model the actions of an intelligent adversary.

To address these concerns, we convert the DHS bioterrorist event tree to a bioterrorist decision tree by changing terrorist decisions to decision nodes, removing two nodes that are problematic and unnecessary, dramatically reducing the complexity by assessing probabilities for each arc for each event instead of probability distributions for each arc for each event. In addition, we describe several alternatives for consequence modeling including separate and aggregated consequences.

MANY BTRA MODELING ALTERNATIVES EXIST

Several risk analysis modeling decisions must be made to provide effective and efficient risk analyses that support national homeland security decision-makers. Figure D.1 is a strategy generation table (Parnell, Driscoll, and Henderson, 2008) used to identify possible modeling decisions. The column titles of Figure D.1 identify some of the most important modeling decisions. The analysis responsiveness (model run time) determines the flexibility of the model and the usefulness to support risk assessment and risk management decision making. The model's transparency increases the understanding and credibility of the model to stakeholders and decision makers. The assumed time period significantly impacts the data collection. The longer the time period, the more challenging it will be to provide credible data assessments. The next three columns (terrorist decisions, U.S. decisions, and uncertain events) are the decisions and events that must be modeled. The types of consequences are major modeling decisions since models will need to be developed for each type of consequence. Finally, the consequences can be modeled individually or combined. Combining enables an integrated assessment but takes more modeling and analysis to credibly combine the consequences.

The columns below the modeling decisions identify several possible techniques for each modeling decision. For example, analysis responsiveness can be real-time, hours, days, weeks, or months. Years are possible but probably not very useful. Using the strategy generation table, we can shade one (or more) box(es) in each column to describe or develop a BTRA modeling alternative. Figure D.1 describes the BTRA of 2006 and Figure D.2 describes the Bioterrorist Decision Model developed in this appendix.

The shading in Figure D.1 shows the committee's understanding of the 2006 BTRA modeling. Battelle developed its own software instead of usually commercially available software to perform the event tree analysis. Due to the complexity, the BTRA model runs in days and requires special software and specially trained analysts to perform the analysis. Some sensitivity analysis capability has been developed and performed. The BTRA model is not transparent. The model is very complex and uses a mixture of best available existing models and new, unvalidated models. The first event in the BTRA event tree is the frequency of attacks. This approach requires specification of a time period and the prediction of the number of attacks with each agent. BTRA event tree models terrorist decisions, U.S. decisions, and uncertain events as probabilities. The methodology greatly increases its complexity and data requirements by assessing probability distributions on each branch of the event tree. The primary consequence modeling was on mortality but some modeling of morbidity and economics was done. The consequences were analyzed individually and not combined.

Analysis Responsiveness (Run-Time)	Model Transparency	Time Period	Terrorist Decisions	U.S. Decisions	Uncertain Events	Consequences	Combining Consequences[a]
Real-time (Minutes)	Transparent, simple models tailored to available data	Time until first attack	Scenarios	Scenarios	Not modeled	Mortality	Analyzed individually and not combined
Hours	Transparent using metamodels developed for best available national models	Fixed time period with potential for multiple attacks	Probability distributions	Probability distributions	Deterministic (parameter)	Morbidity	Converted to dollars
Days	Black box with models that are mixture of best available and unvalidated models	Multiple attacks in a specified time period	Decision made to maximize some objective(s)	Decision made to maximize some objective(s)	Probability distribution	Economic	Combined with multiattribute value function
Weeks	Black box with unvalidated, unverified, and unaccredited models	Multiple attacks in an unspecified time period	Game theory models		Probability distributions on probabilities	Psychological	Combined with multiattribute utility function
Months	Distributed modeling using best available national models	Not applicable	Attacker-defender models		Not applicable	Environmental	Not applicable

[a]Kirkwood (1997) discusses the technical assumptions for multiattribute value and utility functions.

FIGURE D.1 BTRA modeling alternatives. This figure provides a bioterrorism risk assessment modeling alternative generation table (Parnell, Driscoll, and Henderson, 2008) to help identify the BTRA modeling alternatives available to DHS. The column headings are the modeling decisions that must be made by DHS. The column cells identify the modeling techniques we considered for each modeling decision. The gray shading depicts the committee's understanding of 2006 BTRA methodology.

USING DECISION ANALYSIS TO ANALYZE THE TERRORIST'S ATTACK DECISION

Based on the committee's assessment, several improvements are needed. First and foremost, the methodology must consider the terrorist as an intelligent adversary that will select the best attack strategy to maximize their strategic objectives. Second, the methodology must be transparent. A key goal should be the use of commercially available software that has built-in sensitivity analysis features to improve understanding and transparency. The method should eliminate unnecessary complexity and demands for data that will have no meaning if one bioterrorism attack is made on the United States, e.g., the attack frequency for each agent. Finally, the methodology should be easily modified to support the analysis of risk management alternatives.

Decision analysis offers the potential to make many of the improvements we have discussed. Decision analysis is closely related to probabilistic risk analysis (Paté-Cornell and Dillon, 2006). Single objective decision analysis with decision trees has been used since 1968 (Raiffa, 1968; Clemen, 1996). Multiple objective decision analysis has been used since 1976 (Keeney and Raiffa, 1976; Kirkwood, 1997). Maxwell (2006) summarizes the large selection of commercially available decision and risk analysis software.

Figure D.2 uses the format of Figure D.1 and shows the modeling techniques that would be used in a decision analysis method. The darker shaded cells define one potential decision analysis method used to maximize the achievement of terrorist objectives. The lighter shaded cells describe alternative decision analysis methods. The goal would be to use commercially available tools and keep the models small enough to have reasonable run times. Using commercially available software helps make the models transparent and allows the use of standard decision analysis and sensitivity analysis that provide insights and improve transparency. The decision tree would model the terrorist's decision to use biological agents to achieve his or her strategic objectives by maximizing consequences to the United States. All of the terrorist decisions would be modeled as decision nodes.

Analysis Responsiveness (Run-Time)	Model Transparency	Time Period	Terrorist Decisions	U.S. Decisions	Uncertain Events	Consequences	Combining Consequences
Real-time (Minutes)	Transparent, simple models tailored to available data	Time until first attack	Scenarios	Scenarios	Not modeled	Mortality	Analyzed individually and not combined
Hours	Transparent using metamodels developed for best available national models	Fixed time period with potential for multiple attacks	Probability distributions	Probability distributions	Deterministic (parameter)	Morbidity	Converted to dollars
Days	Black box with models that are mixture of best available and unvalidated models	Multiple attacks in a specified time period	Decision made to maximize some objective(s)	Decision made to maximize some objective(s)	Probability distribution	Economic	Combined with multiattribute value function
Weeks	Black box with unvalidated, unverified, and unaccredited models	Multiple attacks in an unspecified time period	Game theory models		Probability distributions on probabilities	Psychological	Combined with multiattribute utility function
Months	Distributed modeling using best available national models	Not applicable	Attacker-defender models		Not applicable	Environmental	Not applicable

FIGURE D.2 BTRA modeling using decision analysis. This figure provides an alternative generation table developed in Figure D.1. However, instead of showing the 2006 BTRA modeling alternative, the dark gray shading highlights a decision analysis method for BTRA. The light gray shading identifies possible variations to the proposed decision analysis methodology. For example, instead of combining the consequences using a multiattribute value model, the consequences could be analyzed individually and not combined or be converted to dollars.

Since they are uncertain to the terrorists, U.S. decisions (e.g., interdiction) and uncertain events (e.g., detection) would be modeled using probability distributions. Any of the consequences that have credible models could be used. Decision trees can be used to find the terrorist strategy (a sequential set of decisions) that maximizes the terrorist objectives by averaging out and rolling back the decision tree. The decision tree can be solved multiple times for each single objective or can be solved once with combined consequences (Parnell, 2007). There are at least three ways of combining the consequences: converting each consequence to dollars, using a multiple attribute value model to normalize and weight the consequences, or using a multiple attribute utility model to normalize and weight the consequences. Each of the techniques has different assumptions and data requirements. All have been used on major national studies.

AN ILLUSTRATIVE BIOTERRORIST DECISION MODEL USING DECISION TREES

The 18 node event tree (with consequences) could be simplified especially if credible data are not available from subject matter experts. However, in order to use as much as possible of the existing 2006 BTRA event tree method, we directly converted the event tree to a decision tree. Using a format similar to Figure 3.4 in Chapter 3 of this report, Figure D.3 lists one possible set of assumptions that could be used to convert the DHS event tree to the bioterrorist decision tree. The figure adds new node numbers, type of node, rationale, average branches, and probability distributions to be assessed. The phases are the same but are not included due to space limitations on the page.

Several assumptions were made in Figure D.3. First, the old nodes numbers 1 (frequency of attack) and 16 (potential for multiple attacks) were deleted for the reasons discussed above. Second, we converted all terrorism decisions to decision nodes.[1] That left six chance nodes: four interdiction nodes, one detection node, and one consequence node. Each of these would be uncertain to the bioterrorist. Third,

[1]While agent selection is an obvious decision, some of the later decisions could be modeled as uncertain nodes early in the terrorist planning cycle. The actual nodes that would be decision or chance nodes would depend on the knowledge of subject matter experts.

Old Stage No.	New Stage No.	Type of Node	Rationale	Decision/Event	Depends on Events	Max Branches	Average Branches	Paths (cumulative)	Maximize Paths (cumulative)	Probability Distributions to Assess (additive)
1	Deleted	Not Applicable	All probabilities will change after first bioattack.	Frequency of Initiation by Terrorist Group						
3	1	Decision	Terrorists will consider the bioagents they can obtain.	Bioagent Selection		28	28	28	28	0
2	2	Decision	Target will be selected to maximize consequences.	Target Selection	1	8	3	84	224	0
4	3	Decision	Mode will be selected to maximize consequences.	Mode of Dissemination (also determines wet or dry dispersal form)	1, 2	9	3	252	2,016	0
5	4	Decision	Mode will be selected to maximize consequences.	Mode of Agent Acquisition	1	4	4	1,008	8,064	0
6	5	Chance	Can be changed by U.S. actions.	Interdiction during Acquisition	1, 4	2	2	2,016	16,128	112
7	6	Decision	Terrorist selects location.	Location of Production and Processing	1	2	2	4,032	32,256	0
8	7	Decision	Depends on agent.	Mode of Agent Production	1	3	3	12,096	96,768	0
9	8	Decision		Preprocessing and Concentration	1, 2, 3, 7	3	3	36,288	290,304	0
10	9	Decision		Drying and Processing	1, 2, 3	3	3	108,864	870,912	0
11	10	Decision		Additives	1, 2, 3	2	2	217,728	1,741,824	0
12	11	Chance	Can be changed by U.S. actions.	Interdiction During Production and Processing	6	2	2	435,456	3,483,648	56
13	12	Decision	Terrorist decision.	Mode of Transport and Storage	1, 2, 3	3	3	1,306,368	10,450,944	0
14	13	Chance	Depends on U.S. actions.	Interdiction During Transport and Storage	6	2	2	2,612,736	20,901,888	56
15	14	Chance	Depends on U.S. actions.	Interdiction During Attack		2	2	5,225,472	41,803,776	1
16		Not Applicable	Terrorist can always do multiple attacks.	Potential for Multiple Attacks	1					0
17	15	Chance	Can be changed by U.S. actions.	Event Detection	1, 2, 3	3	3	15,676,416	125,411,328	252

FIGURE D.3 This figure describes one possible set of assumptions that would generate a decision tree that could be solved for a bioterrorist to maximize the consequences of damage to the United States. The figure uses the format of Figure 3.4 in Chapter 3 of this report and adds new node numbers, type of node, rationale, average branches, and probability distributions to be assessed. All terrorist decisions are converted to decision nodes.

we added the consequence model to the decision tree as the end node. In decision analysis software, this would be implemented using an equation in the end node that uses scenario parameters common to all agents and parameters (agent decision and chance node outcomes) that depend on the path through the decision tree. If the consequences are not combined, a decision tree would be created for each consequence using a different consequence model.

THE BIOTERRORIST DECISION MODEL CAN PROVIDE RISK ASSESSMENT RESULTS AND SENSITIVITY ANALYSIS

The decision analysis model that we have described would identify the terrorist's best strategy to maximize the consequences of an attack. Senior decision makers and stakeholders would be provided a one to *n* list of the agents that have the potential to create the most harm to the United

States. Since decision analysis also calculates the cumulative consequence distribution for each strategy, absolute risk could easily be displayed for each agent.

Decision analysis models are transparent. Commercial decision analysis tools provide a range of powerful sensitivity analysis tools (Clemen, 1996) to increase understanding and improve credibility. The model can be quickly resolved if any stakeholder provides an alternative set of data assumptions. Sensitivity analysis bar charts (Tornado diagrams) can be used to show the most significant data assumptions. Value of information calculations can be performed to find out what uncertainties have the most impact on the agent risk.

THE BIOTERRORIST DECISION MODEL ALSO SUPPORTS RISK MANAGEMENT DECISION MAKING

So far we have focused on the use of decision analysis as a modeling framework to support bioterrorism risk assessments. The Bioterrorist Decision Model would provide the baseline risk for the bioagents analyzed. Since the model can be run quickly, it could be a very useful tool to support DHS risk management decision making.

The bioterrorism risk is impacted by the U.S. ability to reduce the threat (prevent an attack or interdict an attack in progress), reduce the nation's vulnerabilities, and mitigate the consequences given that an attack has occurred. Government agencies, including the intelligence community, the Department of Homeland Security, and the Department of Health and Human Services, expend significant resources each year to increase security against attacks on our nation, including bioterrorist attacks. In the Bioterrorist Decision Model, U.S. capabilities are reflected in the probabilities assigned to the uncertain nodes (the interdiction, detection, and consequence nodes). To assess the risk reduction of risk management alteratives we can modify the model to change the probabilities for each risk management alternative or set of alternatives. Due to the complexities of risk assessment mentioned in Chapter 2 of this report, the results may be initially non-intuitive. For example, a large reduction in the consequences of the highest-risk bioagent may not have a large reduction in overall risk since the second-highest-agent consequences might not be affected. In some cases, we would have to consider sets of alternatives since, in general, the risk reduction would not be additive. Some risk management alternatives may be synergistic (impact greater than the sum of their individual benefits) or complementary (impact less than the sum of their individual benefits).

INSIGHTS FROM THE BIOTERRORIST DECISION MODEL APPROACH

There are several important insights from the analysis presented in this appendix. First, converting the event tree to a decision tree greatly simplifies the probability assessment tasks. Second, the decision tree should allow the tree to be solved using commercially available software using complete enumeration or Monte Carlo simulation. Third, the new challenge is how to develop consequence models that use the decision parameters in the decision tree that will allow for rapid evaluation of the decision tree for each path. Fourth, further opportunities exist to simplify the decision tree. For example, if a decision does not impact the consequences, it can be removed from the decision tree.

THE BIOTERRORIST DECISION MODEL EFFECTIVELY ADDRESSES THE FUNDAMENTAL CONCERNS OF THE BTRA OF 2006

In the introduction we listed the most fundamental concerns with the 2006 BTRA methodology: not considering intelligent adversary decision making, huge data demands, more complexity than the available data support, lack of transparency for decision makers/stakeholders (see Chapter 3), and lack of a clear linkage to DHS risk management decision making. The Bioterrorist Decision Model effectively addresses each of these concerns.

The Bioterrorist Decision Model solves the problem of modeling an intelligent adversary by selecting the bioagents that will maximize the objectives of the terrorists. The model greatly reduces the huge data demands by converting terrorist decisions to decision nodes, deleting the two most problematic nodes—frequency of attack and multiple attacks—and not using probability distributions for each arc on each node. Finally, the model improves transparency by using commercially available software with built-in sensitivity analysis capabilities.

REFERENCES

Clemen, R. 1996. *Making Hard Decisions*, 2nd edition. Belmont, Calif.: Duxbury Press.

Dillon-Merrill, R.L., G.S. Parnell, and D.L. Buckshaw. 2007. "Logic Trees: Fault, Success, Attack, Event, Probability, and Decision Trees." In John G. Voeller (ed.), *Wiley Handbook of Science and Technology for Homeland Security*. Hoboken, N.J.: Wiley and Sons. Forthcoming.

Keeney, R.L. 1992. *Value-Focused Thinking: A Path to Creative Decision-making*. Cambridge, Mass.: Harvard University Press.

Keeney, R.L., and H. Raiffa. 1976. *Decision Making with Multiple Objectives Preferences and Value Tradeoffs*. New York: Wiley.

Kirkwood, C.W. 1997. *Strategic Decision Making: Multiobjective Decision Analysis with Spreadsheets*. Belmont, Calif.: Duxbury Press.

Maxwell, D.T. 2006. "Improving Hard Decisions." *OR/MS Today*, pp. 51-61. [Biannual survey of decision analysis software]

Parnell, G.S. 2007. "Multi-objective Decision Analysis." In John G. Voeller (ed.), *Wiley Handbook of Science and Technology for Homeland Security*. Hoboken, N.J.: Wiley & Sons. Forthcoming.

Parnell, G.S., P.J. Driscoll, and D.L. Henderson (eds.). 2008. *Decision Making for Systems Engineering and Management*. Wiley Series in Systems Engineering, Andrew P. Sage (ed.). Hoboken, N.J.: Wiley and Sons.

Paté-Cornell, E.E., and R.L. Dillon. 2006. "The Respective Roles of Risk and Decision Analysis in Decision Support." *Decision Analysis* 3(4):220-232.

Raiffa, H. 1968. *Decision Analysis: Introductory Lectures on Choices Under Uncertainty*. Boston, Mass.: Addison-Wesley.

Appendix E

Optimizing Department of Homeland Security Defense Investments: Applying Defender-Attacker (-Defender) Optimization to Terror Risk Assessment and Mitigation

Gerald G. Brown
Distinguished Professor, Department of Operations Research
Naval Postgraduate School, Monterey, California

W. Matthew Carlyle
Associate Professor, Department of Operations Research
Naval Postgraduate School, Monterey, California

R. Kevin Wood
Professor, Department of Operations Research
Naval Postgraduate School, Monterey, California

The U.S. Department of Homeland Security (DHS) is investing billions of dollars to protect us from terrorist attacks and their expected damage (i.e., risk). We present prescriptive optimization models to guide these investments. Our primary goal is to recommend investments in a set of available defense options; each of these options can reduce our vulnerability to terrorist attack, or enable future mitigation actions for particular types of attack. Our models prescribe investments that minimize the maximum risk (i.e., expected damage) to which we are exposed. Our "Defend-Attack-Mitigate risk-minimization model" assumes that terrorist attackers will observe, and react to, any strategic defense investment on the scale required to protect our entire country. We also develop a more general tri-level "Defender-Attacker-Defender risk-minimization model" in which (a) the defender invests strategically in interdiction and/or mitigation options (for example, by inoculating health-care workers, or stockpiling a mix of emergency vaccines), (b) the attacker observes those investments and attacks as effectively as possible, and (c) the defender then optimally deploys the mitigation options that his investments have enabled. We show with simple numerical examples some of the important insights offered by such analysis. As a by-product of our analysis we elicit the optimal attacker behavior that would follow our chosen defensive investment, and therefore we can focus intelligence collection on telltales of the most-likely and most-lethal attacks.

INTRODUCTION

Since September 11, 2001, the U.S. Department of Homeland Security (DHS) has marshaled significant resources to assess the risk to our populace from terrorist attacks of all kinds. The work we report here is directly motivated by just one such risk assessment: pursuant to Homeland Security Presidential Directive 10 (HSPD-10) (The White House, 2004), DHS has conducted an extensive bioterrorism risk-assessment exercise, referred to here as the Biological Threat Risk Assessment (BTRA) (DHS, 2006). BTRA estimates risks of many bioterror attack possibilities, and classifies a list of particular bioterror agents as *most-*, *intermediate-*, and *least-threatening*.

The BTRA risk assessment depends upon subject-matter experts (SMEs) advising, with perfect knowledge, the probability that the "attacker" (terrorist or terrorist group), or "defender" (the federal government), will choose some particular option at each stage of an 18-stage probability risk assessment tree.

We contend that representing intelligent adversarial decisions with static probabilities elicited from SMEs is an untenable paradigm: Not only can experts make mistakes, but static probabilities make no sense when the attacker can observe and react, dynamically, to any earlier decisions made by the defender.

We also hold that the business of DHS lies not just in assessing risks, but also in wisely guiding investments of our nation's wealth to reduce these risks. These are strategic *decisions* that must be made now, in a deliberative fashion.

Here, we try to adopt the same problem context as BTRA to recoup its estimable investment in risk modeling. But, we distinguish between (a) strategic investment decisions that DHS makes that are visible to terrorists, (b) the decision a terrorist makes to attempt an attack and, finally, (c) the after-attack mitigation efforts that prudent DHS investments will have enabled.

Our work applies equally well to any category of threat that concerns DHS enough to warrant investments so significant they cannot be hidden from our taxpayers, and thus not from terrorists, either. Such threats cover biological, tadioactive, chemical, and conventional attacks on our infrastructure and citizens, as well as sealing our borders against illegal immigration, and a host of military topics.

The modeling presented here has been motivated and validated by more than one hundred worldwide infrastructure vulnerability analyses conducted since 9/11 by the military-officer students and the faculty of the Naval Postgraduate School (Brown et al., 2005a, 2006a). Some of these studies have been developed into complete decision-support systems:

- Salmerón et al. (2004) have received DHS and Department of Energy support to create the Vulnerability of Electric Grids Analyzer (VEGA), a highly detailed, optimization-based decision-support system. VEGA can evaluate, on a laptop computer, the vulnerability and optimal defense of electrical generation and distribution systems in the United States, where risk is measured as expected unserved demand for energy during any repair-and-recovery period.
- We have developed a decision-support system to advise policy makers regarding the interdiction of a proliferator's industrial project to produce a first batch of nuclear weapons (Brown et al., 2006b, 2007).
- The U.S. Navy has developed a decision-support system to optimally pre-position sensor and defensive interceptor platforms to protect against a theater ballistic missile attack (Brown et al., 2005b).

The message here is that, with experience, we have gained confidence that these new mathematical methods produce results that exhibit the right level of detail, solve the right decision problems, and convey useful advice and insight to policy makers. Such capabilities have not been available before.

THE MODEL, "MXM"

The Biological Threat Risk Assessment (BTRA) uses a descriptive model. Our focus is prescriptive, rather than descriptive: our models suggest prudent investment and mitigation plans for biodefense, and we strive to provide a realistic representation of the attack decisions made by an intelligent adversary.

As the *defender*, we seek to allocate a limited budget among biodefense investment options to form a defense strategy that minimizes the maximum *risk* from the actions of a terrorist *attacker*. We might define risk as the expected number of fatalities, or as the expected 95th percentile of fatalities, or as any other gauge that appeals. Risk is a somewhat ambiguous term when used to discuss our bilateral view of conflict between intelligent adversaries, so we hereafter substitute "expected damage to the defender." We assume that an intelligent adversary will attempt to inflict maximum expected damage. The following, simplified model minimizes a reasonable upper bound on expected damage; we discuss generalizations later.

- **Indices**

$d \in D$	defense strategy, e.g., stockpile vaccines A and B, but not C
$a \in A$	attack alternative, e.g., release infectious agent V
$m \in M$	after-attack, mitigation activity, e.g., distribute vaccine A
$m \in M_d$	mitigation activities enabled by defense option d, e.g., distribute vaccine A, distribute vaccine B
$d \in D_m$	defense strategies that enable mitigation activity m
$k \in K$	resource types used by mitigation activities, e.g., aircraft for distributing vaccine, personnel for administering vaccine

- **Data**

$damage_{d,a}$	expected damage if defense strategy d and attack alternative a are chosen, given no mitigation
$mitigate_{d,a,m}$	expected damage reduction of after-attack mitigation effort m, given investment strategy d and attack a (assumes additive reduction and $\sum_m mitigation_{d,a,m} \leq damage_{d,a}$)
$r_{k,d}$	total mitigation resource of type k available if defense strategy d is chosen
$q_{k,d,m}$	consumption of mitigation resource k provided by defense option d for mitigation activity m

- **Decision Variables**

w_d	1 if defense strategy *d* chosen, otherwise 0
x_a	probability attacker chooses attack alternative *a* ($0 \leq x_a \leq 1$)
$y_{d,m}$	fraction of defense strategy *d* effort devoted to mitigation activity type *m*

- **Formulation: MIN-MAX-MIN (MXM) (Defender-Attacker-Mitigator)**

$$z^* = \min_{w_d} \max_{x_a} \min_{y_{d,m}} \sum_{d,a} damage_{d,a} w_d x_a - \sum_{d,a,m} mitigate_{d,a,m} x_a y_{d,m} \quad \text{(D0)}$$

$$\sum_d w_d = 1 \quad \text{(D1)}$$

$$\sum_a x_a = 1 \quad \text{(A1)}$$

$$\sum_{d,m} q_{k,d,m} y_{d,m} \leq \sum_d r_{k,d} w_d \quad \forall k \in K \quad \text{(M1)}$$

$$y_{d,m} \leq w_d \quad \forall d \in D, m \in M_d \quad \text{(M2)}$$

$$w_d \in \{0,1\}, x_a \geq 0, y_{d,m} \geq 0 \quad \forall d \in D, a \in A, m \in M_d$$

Description

The order of appearance of the operators, min, followed by max, followed by min, in the objective function (D0) represents the sequential nature of the decisions we are modeling, from the outside to the inside. The coefficient $damage_{d,a}$ in the objective accounts for any interdiction effects that strategy *d* has on attack *a*, effects that are independent of any mitigation activities. (For example, vaccinating emergency and health-care providers falls under the category of "interdiction": after an attack, no follow-up mitigation efforts apply to this vaccination.) The right-most minimization term, over $y_{d,m}$, subtracts from expected damage if a mitigating effort has been enabled by the defense plan, and if some amount of that mitigation is applied. For simplicity of exposition, we assume that mitigation results are additive and restricted to sum to some value not exceeding total expected damage. (See the definition of $mitigate_{d,a,m}$.) Constraint (D1) simply limits the defender to choosing one defense strategy. Constraint (A1) limits the attacker to choosing a mixed attack strategy, which of course admits a pure attack as well. Constraints (M1) are joint resource constraints on mitigation efforts; constraints (M2) stipulate that mitigation efforts are permitted only if the enabling defense strategy has been chosen. Constraints (M1) subsume those of type (M2), but we keep these separate for later clarity. The attack variables, x_a, and the mitigation variables, $y_{d,m}$, are continuous. If the attacker variables are restricted to be integer (for instance, they might be binary variables indicating whether or not the terrorists decide to fully develop and deploy a particular pathogen in an attack), then the resulting analysis becomes significantly more complicated than that which we present here. Although dealing with bioterrorist attacks might be most naturally modeled using integer attacker variables, our model with continuous attack (y_a) variables will at least provide a conservative estimate of the defender's objective; i.e., the attacker's abilities to inflict damage are over-estimated by our model.

Discussion of MXM

Figure E.1 depicts a tree showing the sequential actions of the defender (selecting a defense strategy), the attacker (choosing attack alternatives), and the defender (mitigating damage with resources put in place by the defense strategy). (We use the generic term "tree" to represent the sequence of defender and attacker decisions we model. The "decision tree" of Raiffa [1968] pits a single decision maker against Mother Nature, while here we have two opponents trying to shape an outcome governed by Mother Nature. The term "game tree" [Kuhn, 1953] is a more appropriate term for our bioterror situation.) Each defense strategy has an immediate effect on the maximum damage of any attack, reflected in $damage_{d,a}$; it can also enable the capability to reduce after-attack damage by as much as $mitigate_{d,a,m}$, if the chosen defense strategy permits a full allocation of mitigation resources to mitigation action *m*. Given a fixed defense strategy, we assume the attacker will first observe this strategy and then respond with a mixed strategy over the set of possible attacks. As we have said, this might be a relaxation of the original optimization problem faced by the attacker, and therefore grants him or her more attack capability than the attacker really has in this sequential decision-making. In general we cannot tell how weak this relaxation is, but for specific cases (especially those with a moderate number

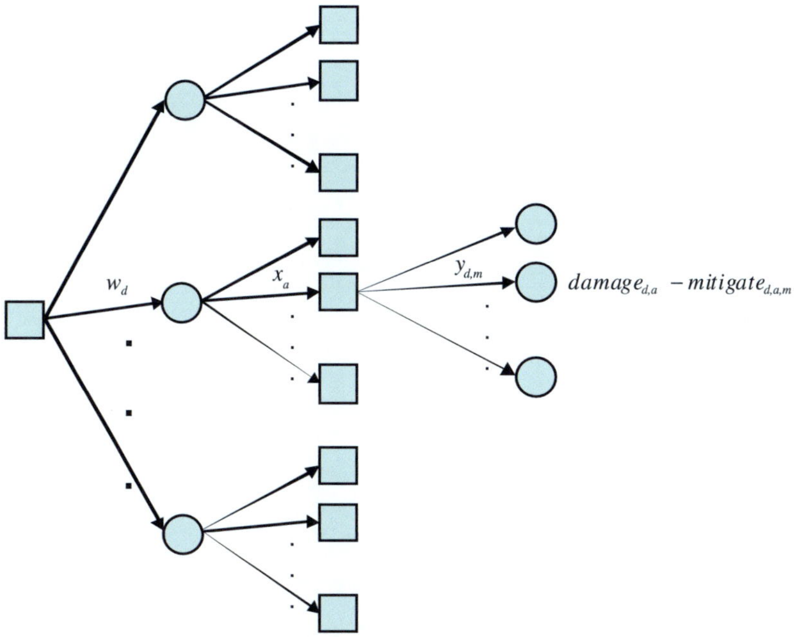

FIGURE E.1 This tree depicts, left-to-right, a leading defense strategy choice w_d, consisting of component defense investment options, and visible to an attacker, followed by attack alternative choice(s) x_a that (each) inflict expected damage $damage_{d,a}$. Square nodes indicate defender decisions, and circle nodes indicate attacker decisions. We only illustrate a mitigation subtree ($y_{d,m}$ decisions) for one (w_d, x_a) pair. For a given defense strategy w_d=1, the optimization recommends a mixed attack strategy for the attacker and a mixed mitigation response $y_{d,m}$ from the defender. The defense strategy establishes all mitigation resources that can be used after an attack. That strategy is seen by the attacker when he or she develops the attack plan. Enabled mitigation resources can reduce expected damage through $-mitigate_{d,a,m} x_a y_{d,m}$. (Our conservative model does not allow the defender to observe the precise type of attack, however, so the mitigation response may not be optimal.)

of feasible attacker decisions) we can use enumeration to bound the effect of this relaxation on the optimal objective function value.

A "mixed attack strategy" means that the optimal attacker decision includes multiple attacks and then we choose mitigation responses, and this results in some damage that can only be estimated, and some part of that estimation can involve an expectation. (For example, the damage could involve an expectation taken over a probability distribution for the time between when an attack is launched to when it is discovered.) Thus, integrating damage over one or more probability distributions yields an objective function that measures "expected damage."

Solving MXM

Temporarily fixing $w = \hat{w}$ in MXM, we take the linear-programming dual (hereafter referred to simply as "the dual") of the innermost minimizing linear program, using dual variables α_k for constraints (M1), and $\beta_{d,m}$ for constraints (M2). This converts the inner "max-min problem" into a "max-max problem," which is a simple maximization:

- **Formulation: MAX-ATTACKER-LP ($\hat{w}$)**

$$z_{\max} = \max_{x,\alpha,\beta} \quad \sum_{d,a} damage_{d,a} \hat{w}_d x_a - \sum_k r_{k,d} \alpha_k - \sum_{d,m \in M_d} \hat{w}_d \beta_{d,m}$$

$$\text{s.t.} \quad \sum_a x_a = 1 \qquad \text{(A1)}$$

$$\sum_k q_{k,d,m} \alpha_k + \beta_{d,m} \ge \sum_a mitigate_{d,a,m} x_a \qquad \forall d \in D, m \in M_d \qquad \text{(DM1)}$$

$$\alpha_k \ge 0 \qquad \forall k \in K$$

$$\beta_{d,m} \ge 0 \qquad \forall d \in D, m \in M_d$$

Now, leaving $w = \hat{w}$ as shown in MAX-ATTACKER-LP, we take the dual of this linear program, using dual variables $\mathfrak{R}$ for constraint (A1) and $y_{d,m}$ for constraints (DM1), and then release w to vary as before, to achieve the following integer linear program which is essentially equivalent to MXM:

- **Formulation: MIN-ILP (Defender-Attacker-Mitigator)**

$$z_{\min} = \min_{\Re, w_d, y_{d,m}} \Re \qquad \text{(DILP0)}$$

$$\text{s.t.} \quad \Re \geq \sum_d damage_{d,a} w_d - \sum_{d,m} mitigate_{d,a,m} y_{d,m} \quad \forall a \in A \qquad \text{(DILP1)}$$

$$\sum_d w_d = 1 \qquad \text{(D1)}$$

$$\sum_{d,m} q_{k,d,m} y_{d,m} \leq \sum_d r_{k,d} w_d \quad \forall k \in K \qquad \text{(M1)}$$

$$y_{d,m} \leq w_d \quad \forall d \in D, m \in M_d \qquad \text{(M2)}$$

$$w_d \in \{0,1\}, y_{d,m} \geq 0 \quad \forall d \in D, m \in M_d$$

The optimal solution to MIN-ILP prescribes among other things a choice for the defense strategy, w^*, to be implemented immediately by the defender, before an attack occurs. Given optimal incumbent solution w^*, we recover the attacker's optimal strategy x^* by solving MAX-ATTACKER-LP(w^*).

A Numerical Example of MXM

We provide a small numerical example to illustrate the features of MXM.

We introduce a number of *defensive investment options,* programs that can be composed in groups into *defense strategies*. Table E.1 displays defensive investment options and costs.

In our example, the defensive investment options are denoted "i01," "i02," and "i03." From this set, policy makers have determined 6 combinations that comprise the subset of admissible defense strategies whose implementation will depend on the available budget; see Table E.2. Table E.3 displays expected damage resulting from each defense strategy and each attack alternative, i.e., the terms $damage_{d,a}$.

Figure E.2 illustrates the generic relationship relating investment options to the ability to reduce expected damage from any terrorist attack before it is carried out, and/or mitigate damage after an attack occurs. This is a complicated function, neither convex nor concave, but our sampling of representative points can be used to represent this in characterizing component investment options in defense strategies.

Damage estimates in Table E.3 include any synergies among or interference between component investment options in each defense strategy preparing for each attack. This is key. BTRA makes a point of such dependencies, and we represent these in complete, realistic detail here.

Table E.4 represents estimated mitigation capabilities. These mitigation estimates correspond to a single, "full-strength" mitigation effort being applied to a single attack alternative. If the attacker chooses a mixed attack strategy, we may need to spread mitigation effort across multiple activities, reducing the expected effectiveness of each activity accordingly.

The choice of defense strategy is limited by a total budget, which we vary over the integers from 0 to 11. We allow full employment of either mitigation effort, or any convex combination of them.

Because the defender is minimizing the optimal objective function value of a *maximization* problem, the optimal

TABLE E.1 Defensive investment options and costs.

i	$cost_i$
i01	2
i02	3
i03	5

For example, option "i03" costs 5. Total budget, logical, and perhaps political considerations will limit the combinations of these options that can comprise admissible defense strategies

TABLE E.2 Defensive investment options in each potential defense strategy.

		Investment options		
		i01	i02	i03
Defensive strategies	d00			
	d01	x		
	d02		x	
	d03			x
	d04	x	x	
	d05		x	x

Strategy "d00" makes no investment at all. Defense strategy "d05" includes investment options "i02" and "i03." Logical, political, or other considerations preclude some of the strategies, for example, {"i01," "i03"}. The total available budget, not yet specified, can also preclude certain strategies. For instance, {"i02" and "i03"} cannot be selected if the total budget is less than 8.

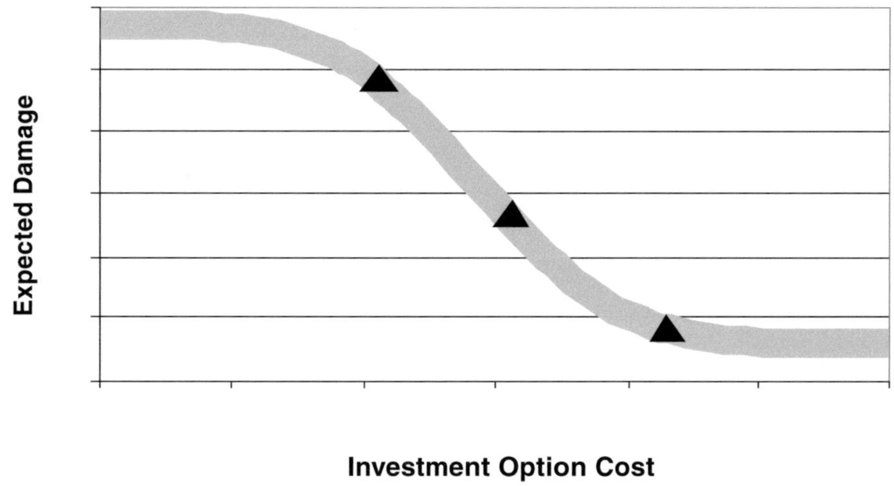

FIGURE E.2 The purpose of Department of Homeland Security defensive investment options is to reduce expected damage before an attack occurs, and/or allow mitigation of expected damage after one occurs. The generic relationship illustrated here conjectures little to no effect at low investment levels, followed by increased effectiveness, and eventually leveling off with diminishing returns. The triangles represent points we might use as alternate investment options to adequately represent the entire function.

TABLE E.3 Expected damage resulting from each defense strategy (row) and each attack alternative (column), accounting for interdiction but not mitigation.

	a01	a02	a03
d00	10	10	10
d01	10	5	7
d02	6	8	7
d03	6	6	6
d04	4	3	5
d05	5	5	4

(This table gives the values for $damage_{d,a}$ for MXM. We use integral data to permit reproduction of our results.)

TABLE E.4 (A, left; B, right) Maximum expected damage reduction from a mitigation activity enabled (prior to an attack) by a defense strategy (and applied after an attack).

m = m1	a01	a02	a03
d00	0	0	0
d01	1	0	0
d02	0	1	1
d03	0	0	1
d04	1	1	①
d05	0	1	1

m = m2	a01	a02	a03
d00	0	0	0
d01	1	0	0
d02	0	2	0
d03	0	0	1
d04	0	1	②
d05	0	0	2

These tables specify $mitigate_{d,a,m}$ for MXM, for each of two mitigation options (Table E.4.A, "m = m1," and Table E.4.B, "m = m2"), for each combination of defense and attack. For example, with defense option "d04" and attack "a03," if we choose mitigation "m = m1" we reduce the damage by one unit, but if we choose mitigation "m = m2" we reduce the expected damage by two units (circled values).

solution invests to reduce the expected damage, given future mitigation capability, of the most-threatening mixed attack. This requires that the defender invest in a defense strategy that enables him or her to mitigate several very-damaging attacks, and not just the worst one.

Figure E.3 shows minimized maximum expected damage as a function of total defense budget, and Table E.5 summarizes the solutions for each budget break-point. For instance, with a budget of 3, the optimal defense plan in MXM is to choose defense option "d02." The terrorists' optimal attack is a mixed strategy, with a probability of 0.50 of choosing "a02" and probability 0.50 of choosing "a03." The resulting expected damage, after mitigation, is 6.5. Analysis of this simple case reveals that we have optimally allocated our mitigation effort among the two worst attacks, reducing the expected damage in each attack to the same value, 6.5. We can do no better than this, given our conservative approximation.

Generalizing Beyond Tri-level Decision Problems

The DHS biological threat risk assessment (BTRA) consists of an 18-stage probability risk assessment tree, where each decision has been replaced by an a priori probability, as shown in Chapter 3 of this report. In the case of the each opponent, these probabilities are determined by subject-matter experts assessing how terrorists might make each decision, and how well DHS will do thwarting a bioagent attack at some intermediate stage of its development.

We could instead model the BTRA as a 19-stage defender-attacker-defender model, with a new stage zero describing how DHS can invest in strategic biological defense

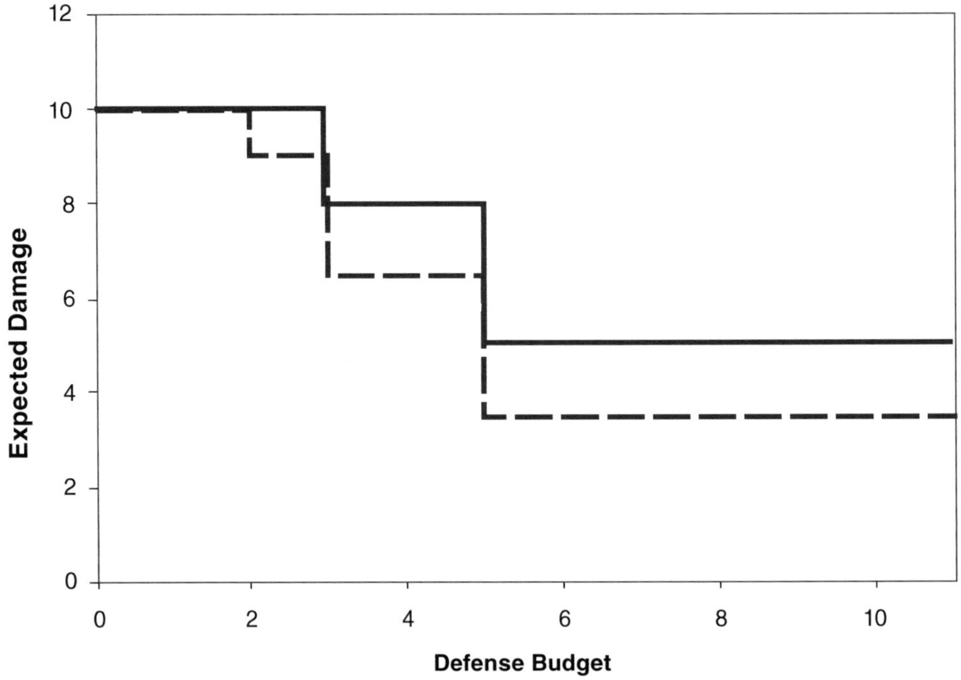

FIGURE E.3 Expected damage as a function of defense budget. This display is for policy makers: as we devote more and more defense budget, we achieve less and less expected damage. Because the defender's investment options here are discrete, each improvement appears as a staircase drop as soon as sufficient budget permits some new, improved cohort of investment defense options, i.e., a new defense strategy. The law of diminishing returns is evident: expected damage reduced by each budget dollar decreases as budget increases. Policy makers can usually put their finger on the spot that appeals in an illustration such as this, perhaps based on criteria not part of the underlying modeling. The uppermost, solid line displays the expected damage when all $\text{mitigation}_{d,a,m}$ values are set to zero (i.e., we have no mitigation capability) and only consider the expected damage from adopting a defense strategy, and then suffer the worst-case attack per expected damage in Table E.3. The dashed line illustrates the expected damage from MXM, the tri-level optimization.

strategies, and each of the intermediate stages represented by a set of decision variables that prescribe attacker or defender behavior, and solve a multi-stage defender-attacker-defender(-attacker-...) model to determine optimal stage-zero

TABLE E.5 For each budget just sufficient to afford a new defense strategy, we show the Defender-Attacker solution and expected damage (i.e., for MXM with y = 0), the Defender-Attacker-Defender solution (for MXM) and expected damage.

	MXM with y = **0**			MXM			
Budget	**w**	**x**	*z**	**w**	**x**	**y**	*z**
0	d00	a01	10	d00	a01	—	10
2	d01	a01	10	d01	a01	m01	9
3	d02	a02	8	d02	a02(.50) a03(.50)	m01(.50) m02(.50)	6.5
5	d04	a03	5	d04	a01(.50) a03(.50)	m01(.50) m02(.50)	3.5

For example, with a budget of 3, the optimal defense strategy in MXM is "d02." The terrorists' optimal attack is a mixed strategy, choosing alternative "a02" with probability 0.50, and "a03" with probability 0.50. We anticipate responding accordingly with "m01," the optimal response to "a01," with the same probability (0.50), and similarly with "m02" with probability 0.5. The resulting expected damage, after optimal mitigation in each case, is 6.5.

investment decisions to minimize expected damage assuming each opponent makes the optimal decision at each node of the corresponding tree. To fully represent the sequential nature of these decisions, we would require all decisions (except maybe those in the final stage) to be modeled with integer variables. However, solving such a model for just two stages of integer decisions is difficult.

We do not have the technology to handle three, much less 18, stages of alternating integer decisions. Allowing continuous decision variables in each of the stages except stage zero (our defense decision variables) would again be a relaxation of the restrictions on the attacker, and could, in some cases, yield extremely weak bounds on our defensive capability.

We now show in the case of a two-stage model how this relaxation from integer to continuous variables reduces the sequential decision problem to a simultaneous two-person zero-sum game.

Consider the bi-level, attacker-defender, max-min optimization formulation: $(A_L D_L)$, where the subscript "L" denotes a linear program (i.e., continuous decision variables, and objective and constraints that are linear in those decision variables):

$(A_L D_L)$

$$\begin{array}{lllll} \max\limits_{x}\min\limits_{y} & g^T x + x^T Q y + c^T y & & [\text{dual variables}] & \\ \text{s.t.} & Ax & \le b & [\pi] & (B1) \\ & Dy & \ge d & [\mu] & (B2) \\ & x \ge 0 & & & \\ & y \ge 0 & & & \end{array}$$

$(A_L D_L)$ is a more general version of the model used by Fulkerson and Harding (1977) and Golden (1978) for their work on continuous network interdiction models.

Take the dual of the inner (defender, "y") problem in $(A_L D_L)$:

$(A_L \bar{D}_L)$

$$\begin{array}{llllll} \max\limits_{x,\mu} & g^T x & + d^T \mu & & & \\ \text{s.t.} & Ax & & \le b & [\pi] & (B1) \\ & -Q^T x & D^T \mu & \le c & [y] & (D2) \\ & x \ge 0 & & & & \\ & & \mu \ge 0 & & & \end{array}$$

This is our standard way to convert a "max-min" problem, for which there is no conventional optimization method, into an equivalent "max-max" problem that is nothing more than a conventional linear program.

Now, reverse the order of play in $(A_L D_L)$ to $(D_L A_L)$:

$(D_L A_L)$

$$\begin{array}{lllll} \min\limits_{y}\max\limits_{x} & g^T x + x^T Q y + c^T y & & [\text{dual variables}] & \\ \text{s.t.} & Ax & \le b & [\pi] & (B1) \\ & Dy & \ge d & [\mu] & (B2) \\ & x \ge 0 & & & \\ & y \ge 0 & & & \end{array}$$

This variation on $(A_L D_L)$ is formulated as if the defender makes a decision first.

Take the dual of the inner, attacker, ("x") problem in $(D_L A_L)$:

$(D_L \bar{A}_L)$

$$\begin{array}{llllll} \min\limits_{y,\pi} & b^T \pi & + c^T y & & & \\ \text{s.t.} & A^T \pi & -Qy & \ge g & [x] & (D1) \\ & & Dy & \ge d & [\mu] & (B2) \\ & \pi \ge 0 & & & & \\ & & y \ge 0 & & & \end{array}$$

This formulation is equivalent to $(D_L A_L)$, and is also a linear program.

We observe that $(A_L \bar{D}_L)$ and $(D_L \bar{A}_L)$ are linear programming duals of each other, and thus (assuming both are feasible) have the same optimal objective-function values, which is the same as the optimal objective value of $(A_L D_L)$. Therefore, the sequence in which the decisions are made (either attacker first, followed by defender, or defender first, followed by attacker) has no impact on the optimal objective-function value.

We have therefore proved the following:

> **Theorem 1:** For any attacker-defender model in the form $(A_L D_L)$, we can exchange the order of decisions without affecting the optimal objective function value.

Theorem 1 is a simple extension of von Neumann's (1928) minimax theorem for polyhedral feasible regions using a proof technique similar to Ville (1938), but using the more modern technology of linear programming duals directly. This exchange argument, along with the observation that any two consecutive decision stages controlled by the same decision maker are equivalent to a single stage (since both stages are either a maximization or both are a minimization over a set of decision variables, this is equivalent to a single maximization, or minimization, over all of those variables simultaneously), can be repeated for any number of consecutive stages with continuous decision variables. The final model obtained in this manner is a simple maximization or minimization problem.

Specifically, if we were to apply this to the 18-stage BTRA model (i.e., the model we would solve for any fixed, known defense decision in stage zero), we would aggregate adjacent attacker stages (and adjacent defender stages, if there are any) and reduce the 18-stage BTRA tree to 8 stages. We would then require that all decision variables be continuous, and then swap adjacent defender-attacker pairs of stages until we obtain a model having all of the attacker decisions in stage 1 and all of the defender decisions in stage 2. This resulting model is equivalent to model $(A_L D_L)$, above, and hence is equivalent to a simultaneous game.

The optimal solution would prescribe mixed strategies for the attacker and defender, eliminating the sequential nature of the real decisions that must be made. In general, the results from such an analysis might not be very accurate, as every relaxation of a block of integer variables to continuous and the subsequent interchange and aggregation of adjacent stages can result in a significant relaxation of attacker restrictions; in some models these approximations could get *significantly* less informative with each additional stage exchanged in this manner.

However, if the sequencing of two adjacent attacker-defender stages is not a critical component of the formulation, then the optimal solution of the relaxation might not be far off from that of the original model. As a simple example, if the attacker chooses which pathogen to load into a truck, and the defender then chooses whether or not to emplace transportation blockades, relaxing the decision variables and exchanging these two stages might not be as significant a relaxation as in a situation where the attacker decides whether or not to release a pathogen, and the defender then chooses whether or not to employ his stockpile of a certain vaccine that can treat the attacker's pathogen. In the former case the blockades will work against the truck regardless of the pathogen chosen, while in the second example committing to use the vaccine before a pathogen is released is clearly a bad idea, and allows the attacker to cause significantly more damage.

How to Generalize BTRA to a Decision Model Prescribing Defense Investments

If we are to leverage the considerable effort that went into the development of the BTRA, we must use the data obtained, and elicit subject-matter-expert input, to develop a two- or three-stage sequential decision model of defensive investments, attacks, and mitigation responses such that the relaxation obtained by allowing continuous attacker variables, as in MXM, is at least a reasonable approximation.

If we are successful in our new modeling effort, then the decisions at each stage except our new stage-zero will be continuous (and, more specifically, interpreted as mixed strategies), but now the values of these mixed-strategy probabilities will be *prescribed* by the optimization model: for the stage under control of the terrorists, these will represent the *worst-case mix of attack decisions the terrorists can devise*; in the mitigation stage, under DHS control, these will represent the best response to each of the attacker's possible decisions in the previous stages.

It is not lost on us that some of the BTRA probabilistic risk assessment tree's probabilities exhibit dependence on the outcomes of some prior stages in the tree. A reformulation to a two- or three-stage sequential decision model would necessarily require some reworking of these data. For brute-force permutation of (potentially aggregated) stages, we could unwind the conditional probabilities with Bayes' theorem (just as DHS already does when it splits the single BTRA tree into 28 independent trees, one for each bioagent, where selection of bioagent is the third terrorist stage in the original tree).

However, we hope to move away from subject-matter-expert (SME) elicitations of highly dependent probabilities as follows. These dependencies are presumably due to the influence of prior stages on the state of the terrorist (or DHS) in terms of exhaustion of limited resources. MXM would explicitly guide strategic defensive investment in stage zero, and subsequently offer all the explicit resource-limiting features of a linear program for all the attacker decisions, and in parallel all the defender's mitigation decisions that consume the mitigation resources provided by stage zero. Linear programming has long been widely applied to planning industrial and military operations that precisely mimic a bioterror-agent production program, or a defense plan.

We recommend eliciting from SMEs an explicit assessment of the resources and capabilities of each opponent, and the way and rate at which various alternate activities would consume these. This is, in fact, the way that the BTRA reports that the SMEs explained their reasoning to support probability assessments. We advise using these technological estimates as explicit inputs, and letting MXM determine attacker mixed-strategy probabilities and expected consequences as outputs. This would be much more transparent modeling, provide better documentation, and be less likely to be influenced by poor SME guesses about high-dimensional decisions governed by complicated resource limitations. This also avoids the current step where SMEs convert capabilities assessments into just a few discrete, qualitative probability classes (e.g., "not likely" = 0.2, "likely" = 0.5, "very likely" = 0.8).

The initial linear integer program and subsequent pair of linear programs afford us a great deal of flexibility and fidelity in describing the actions of each opponent, and we can solve these at very large scale with off-the-shelf optimization software. Also, solutions to such optimization models can be analyzed to discover the "why" as well as the "what" of each plan. Powerful, effective sensitivity and parametric analysis techniques are well known for these optimization models.

We represent defensive investment strategy selection simply, as we think realistic and politically palatable during this early phase of homeland security capital planning. We anticipate that this will eventually mature to more closely resemble classic military capital planning (e.g., Brown et al., 2004).

We present a deterministic model that minimizes the maximum expected risk. If stochastic evaluation proves essential, our model can be used within a simulation. Banks and Anderson (2006) demonstrate such exogenous simulation with a two person, zero-sum game. Tintner (1960) shows this for a linear program. Our integer linear program is amenable to such simulation.

Secrecy in Planning

If, as the defender, we strongly believe that we are able to conceal some of our defensive capability from the attacker, then the transparency of model MXM is likely to be inappropriate for determining optimal defense decisions. Instead, we find ourselves in an *asymmetric* conflict: the attacker and the defender *do not agree on the objective function.* This more general case falls in the domain of bilevel and multilevel programming (see, for example, Candler and Townsley [1982], Bard and Moore [1992], and Migdalas et al. [1998]), and the associated mathematical models are more difficult to solve than those we have presented here.

In an extreme case, for example, we might believe that even though the attacker can observe our strategic defensive investments, he or she is completely unaware of our mitigation capabilities. We could then assume that the attacker will make decisions based only upon the $damage_{d,a}$ values, whereas, given that we are perfectly aware of our mitigation capabilities, we will make our investment decisions based on $damage_{d,a}$–$mitigate_{d,a,m}$ values. This would be formulated as a tri-level integer programming model, the most general versions of which are difficult to solve. However, a straightforward heuristic for solving our problem would solve an attacker-defender version of the problem with no mitigation options (i.e., by fixing $y_{d,m} = 0$), and then choose the optimal mitigation decision for whatever defense and attack decisions are made.

Clearly this can lead to a suboptimal defense investment, especially when there are defense options that do not directly reduce expected damage (i.e., $damage_{d,a}$ might be high for those defenses) but that enable mitigation efforts that are significantly more effective than those available under other defensive investments. We can use the stockpiling of a vaccine as an example; creating the stockpile will not reduce the damage of any attack, but the mitigation activity of distributing the vaccine and inoculating the susceptible population can be extremely effective. In this case, the optimal defense and the resulting worst-case attack damage can differ significantly from the myopic defense. There are other, more effective heuristics for multilevel optimization in the literature, the breadth of which is beyond the scope of this appendix.

In the case where the "secret" objective values maintain the same relative ranking between each pair of feasible defense and attack combinations as discloseded by the "public" objective function, then the optimal defense and resulting worst-case attack do not change. For example, if the mitigation effects $mitigate_{d,a,m}$ are always a fixed percentage of $damage_{d,a}$, then the optimal defensive investments, and the corresponding worst-case attack, will be the same, and the overall expected damage will be reduced by that fixed percentage. In this case (and similar cases, in which the mitigation efforts do not produce drastically different results from each other relative to the defense-and-attack combination they are applied to), it makes no sense to take extreme measures to conceal our mitigation capability. In fact, we should broadcast it widely, in hopes that it will deter attacker efforts.

However, in the case where our mitigation capabilities are much more (or less) effective for one (or a small number of) attacks than for the rest, and this fact fundamentally changes the worst-case attack decision for each of our defense options, then we conjecture that we should conceal this capability to maintain our advantage (or conceal our weakness) for that attack, and hopefully "shape" the attacker's decisions toward the attacks that we are more capable of handling. However, every situation is different, and it is extremely hard to predict what the effect any given "secrecy policy" will have on the optimal outcome, much less on the actual attacker behavior. More research in this area is required.

Solving MXM at Very Large Scale with Decomposition

Although we have solved large attacker-defender models of the same *form* as MXM (Brown et al., 2005b), if instances of MXM become too large to solve using commercial off-the-shelf integer linear programming software, we can use (and have used) a version of Benders decomposition (e.g., Bazaraa, Jarvis, and Sherali, 1990, pp. 366-367) to solve MIN-ILP, with integer stage-zero investment decisions and continuous mitigation decisions in the master problem, and the resulting attacker LP subproblems. Israeli and Wood (2002) explicitly develop such a decomposition for the case of shortest-path network interdiction problems.

We modify MIN-ILP, replacing equations (DILP1) with a set of constraints (DILP-CUTS), and calling the resulting model MIN-ILP-DECOMP($\{\hat{x}^N\}$), where $\{\hat{x}^N\}$ represents the set of all attacker plans from completed decomposition iterations: $\{\hat{x}^N\} \equiv \{\hat{x}^n, n = 1,\ldots,N\}$.

$$\Re \geq \sum_{d,a} damage_{d,a} w_d \hat{x}^n_a - \sum_{d,a,m} mitigate_{d,a,m} \hat{x}^n_a y_{d,m} \qquad n = 1,\ldots,N \qquad \text{(DILP-CUTS)}.$$

The complete decomposition algorithm is as follows:

- **Algorithm DHS-MXM-DECOMP**

  Input: Data for bio-terror defense problem, optimality tolerance $\varepsilon \geq 0$;
  Output: ε-optimal (MXM) defender plan (w^*, y^*);
  1) Initialize best upper bound $z_{UB} \leftarrow \infty$, best lower bound $z_{UB} \leftarrow 0$, define the incumbent, null (MXM) defender plan ($w^* \leftarrow \hat{w}^1 \equiv$ "d00", $y^* \leftarrow y^1 \leftarrow 0$) as the best found so far, and set iteration counter $N \leftarrow 1$;
  2) Subproblem: Using $\hat{w} = \hat{w}^N$, solve the linear program subproblem MAX-ATTACKER-LP ($\hat{w}$) to determine the optimal attack plan $\hat{x}^N$; the bound on the associated total expected target damage is $z_{\max}(\hat{x}^N)$;
  3) If ($z_{UB} > z_{\max}(\hat{x}^N)$) set $z_{UB} \leftarrow z_{\max}(\hat{x}^N)$ and record improved incumbent MXM defender plan $(w^*, y^*) \leftarrow (\hat{w}^N, \hat{y}^N)$;

4) If ($z_{UB} - z_{LB} \leq \varepsilon$) go to End;
5) Given attack plans $\{\hat{x}^N\}$, attempt to solve master problem MIN-ILP-DECOMP($\{\hat{x}^N\}$) to determine an optimal defender plan ($\hat{w}^{N+1}, \hat{y}^{N+1}$). The bound on the total expected target damage is $z_{\min}(\hat{w}, \hat{y})$;
6) If $z_{LB} < z_{\min}(\hat{w}, \hat{y})$ set $z_{LB} \leftarrow z_{\min}(\hat{w}, \hat{y})$;
7) If ($z_{UB} - z_{LB} \leq \varepsilon$) go to End;
8) Set $N \leftarrow N + 1$ and go to step (2) (Subproblem);
9) End: Print "(w^*, y^*) is an ε-optimal (MXM) defender solution," and halt.

The optimal attacker plan x^* can be recovered by solving MAX-ATTACKER-LP(w^*).

Each instance of MAX-ATTACKER-LP($\hat{w}$) is a linear program of a form we expect to be easy to solve even at large scale.

MIN-ILP-DECOMP($\{\hat{x}^N\}$) is easy to solve, but might get more challenging if embellished with too many more linear constraints. For a difficult instance, or at very large scale, we can solve MIN-ILP-DECOMP($\{\hat{x}^N\}$) with an approximate, but very fast heuristic, and our decomposition is still valid.

The iterative behavior of the decomposition is instructive. Set a defense plan, and observe the attack response. Set another defense plan that is robust with respect to the attack response observed, and then observe another attack response. As such iterations continue, the defender learns more about the attacker, and refines his defense plan accordingly. Ultimately, the defender learns enough to declare that his best defense plan is (ε–) optimal against the best possible attacker plan, and attains a mathematical certificate of the quality of his defense preparations. (See Table E.6.)

The decomposition mathematically represents two opposed sets of subject-matter experts: a Blue Team (defender), and Red Team (attacker). The decomposition iterations mathematically mimic a wargame between these opponents, where the defender suffers the disadvantage of not being able to hide the defense strategy, but the players play the game again and again, honing their respective strategies, until neither opponent can improve.

At ultra-large scale, we can nest decompositions. We do not anticipate this will be necessary for this application.

We have implemented MXM and our decomposition algorithm for solving it in GAMS (2007). All model instances have been solved optimally. The complete implementation is available from the authors.

How Do We Get Here from a Descriptive Risk Assessment (e.g., DHS BTRA)?

First, we must recognize and accept that each event-tree path in the BTRA consists almost exclusively of a set of *decisions*—these are *not* random events. There are 18 successive "events" in the National Research Council rendition of BTRA (see Tables E.3 and E.4). From start to finish, we show each event number, using parentheses to distinguish defender actions, and brackets for Mother Nature at the end: the BTRA event sequence is 1-5; (6); 7-11; (12); 13; (14-15); 16; (17); [18]. The first attacker event sequence addresses selection of agent, target method of dissemination, and acquisition; the next attacker sequence involves details of agent production and processing; the following attacker sequence describes transport and storage; and the last estimates repeated attacks. These attacker sequences are interrupted by opportunities for the defender to interdict. The last stage [18] represents Mother Nature influencing consequences. For our purposes, there are merely four alternations from attacker to defender, followed by one truly random event governed by Mother Nature at the end.

Second, we decide how to reckon $damage_{d,a}$ as a function of defense strategy d and attack alternative a. This is not a glib statement, but rather a meta-design guide to return to the foundations of BTRA and critically review the assumptions of sequence-dependence and level of detail.

In theory, this could be achieved by setting a defender option d, and estimating the consequences of this action on BTRA for each pure attacker response. This is no harder than for BTRA, and if we concentrate on estimating $damage_{d,a}$

TABLE E.6 Decomposition iterations reveal learning by opponents. Here, the defender starts with defense strategy "d00" (do nothing), the attacker responds with his most-damaging alternative "a03" inflicting damage 10.

Iteration	Defense Strategy	Lower Bound	Attack Alternative	Upper Bound	
n	MIN-ILP-DECOMP	z_{LB}	MAX-ATTACKER-LP	z_{UB}	Mitigation
1	"d00"	0	"a03"	10	"m01"
2	"d05"	2	"a01"	5	"m02"
3	"d04"	3.5	"a01"(0.5) "a03"(0.5)	3.5	"m01"(0.5) "m02"(0.5)

Subsequent iterations adjust defense strategy based on elicited attacker behavior, until neither opponent can take another turn for any further improvement. Our subject-matter experts (SMEs) are now optimization models. The last iteration yields the same optimal solution as shown in Table E.5. Instead of using a "do-nothing" solution to initialize the algorithm, we can just as easily take any feasible incumbent proposed by any decision maker as our first attempt: the algorithm will evaluate this solution, and then either obtain a certificate of its optimality, or find a better incumbent. This is the distinguishing advantage of viewing these decomposition algorithms as "learning" methods that iteratively improve upon an incumbent, possibly suboptimal, solution.

as a function of defense option d and a more palatable (i.e., unlike BTRA, a less minutely-detailed and less overwhelmingly numerous) set of attack alternatives a, we would create a risk-calculation engine that is at once credible and efficient.

By whatever means, we must estimate $damage_{d,a}$ for each defense option d and each attack alternative a. *If we cannot estimate risks at this fidelity, we have no business doing risk analysis.*

We would prefer to be able to choose a number of defense strategies, rather than just one. But, current risk analysis produces a single damage estimate distribution for each attack scenario. We assume these damage estimates are neither additive nor separable between and among attacks, so we must rely on the simplified risk analysis we have. Accordingly, we endow each defense strategy with the number of defense investment options reflected in each BTRA path.

Our attack alternatives have not specified any particular agent. Our methods can accommodate attacks by classes of agents that include engineered and future agents not yet known.

Solving the tri-level model achieved here isolates an optimal defense strategy, and all its component investment options. Because this optimal strategy dominates every attack by any agent, we have presented an intrinsic risk analysis that highlights the most-critical, achievable defense strategy. We can trivially rule out this best strategy, and solve for the second-best, and so forth. *This renders an explicit, unambiguous prioritization of defense strategies.*

Mere Probabilistic Risk Assessment Is Not Enough

What we are proposing here responds directly to the explicit language of HSPD-10 (The White House, 2004): "the United States requires a continuous, formal process for conducting routine capabilities assessments to guide prioritization of our on-going investments in biodefense-related research, development, planning, and preparedness."

Further, we could not agree more with this: "Successful implementation of our program requires optimizing critical cross-cutting functions" (The White House, 2004).

Recently, HSPD-18 (The White House, 2007) has further clarified our direction: "optimize the investments necessary for medical countermeasures development, and ensure that our activities significantly enhance our domestic and international response and recovery capabilities." Further: "Mitigating illness and preventing death are the principal goals of our medical countermeasure efforts."

Moving beyond mere descriptive risk analysis, we want to address:

> (a) Target threats that have potential for catastrophic impact on our public health and are subject to medical mitigation;
>
> (b) Yield a rapidly deployable and flexible capability to address both existing and evolving threats;
>
> (c) Are part of an integrated weapons of mass destruction consequence management approach informed by current risk assessments of threats, vulnerabilities, and capabilities; and
>
> (d) Include the development of effective, feasible, and pragmatic concepts of operation for responding to and recovering from an attack. (The White House, 2007)

We can see from these policy directives that the highest-level DHS problem is *planning investments*—huge investments—to prepare to *mitigate* the consequences of any attack.

The material presented here follows both the letter and the spirit of this direction.

REFERENCES

Banks, D., and S. Anderson. 2006. "Combining Game Theory and Risk Analysis in Counterterrorism: A Smallpox Example." Pp. 9-22 in A. Wilson, G. Wilson, and D. Olwell (eds.), *Statistical Methods in Counterterrorism*. New York: Springer.

Bard, J., and J. Moore. 1992. "An Algorithm for the Discrete Bilevel Programming Problem." *Naval Research Logistics* 39(3):419-435.

Bazaraa, M.S., J. Jarvis, and H.D. Sherali. 1990. *Linear Programming and Network Flows*. New York: Wiley.

Brown, G., R. Dell, and A. Newman. 2004. "Optimizing Military Capital Planning." *Interfaces* 34(6):415-425.

Brown, G., M. Carlyle, J. Salmerón, and K. Wood. 2005a. "Analyzing the Vulnerability of Critical Infrastructure to Attack, and Planning Defenses." In H. Greenberg and J. Smith (eds.), *Tutorials in Operations Research: Emerging Theory, Methods, and Applications*, Hanover, Md.: Institute for Operations Research and Management Science.

Brown, G., M. Carlyle, D. Diehl, J. Kline, and K. Wood. 2005b. "A Two-Sided Optimization for Theater Ballistic Missile Defense." *Operations Research* 53(5):745-763.

Brown, G., M. Carlyle, J. Salmerón, and K. Wood. 2006a. "Defending Critical Infrastructure." *Interfaces* 36(6):530-544.

Brown, G., M. Carlyle, R. Harney, E. Skroch, and K. Wood. 2006b. "Anatomy of a Project to Produce a First Nuclear Weapon." *Science and Global Security* 14(2/3):163-182.

Brown, G., M. Carlyle, R. Harney, E. Skroch, and K. Wood. 2007. "Interdicting a Nuclear Weapons Project." In review.

Candler, W., and R. Townsley. 1982. "A Linear Two-Level Programming Problem." *Computers and Operations Research* 9(1):59-76.

DHS (Department of Homeland Security). 2006. *Bioterrorism Risk Assessment*. Biological Threat Characterization Center of the National Biodefense Analysis and Countermeasures Center. Fort Detrick, Md.

Fulkerson, D.R., and G.C. Harding. 1977. "Maximizing the Minimum Source-Sink Path Subject to a Budget Constraint." *Mathematical Programming* 13(1):116-118.

GAMS (General Algebraic Modeling System). 2007. "General Algebraic Modeling Language GAMS." Available at http://www.gams.com/. Accessed January 12, 2007.

Golden, B. 1978. "A Problem in Network Interdiction." *Naval Research Logistics Quarterly* 25(4):711-713.

Israeli, E., and R.K. Wood. 2002. "Shortest-Path Network Interdiction." *Networks* 40(2):97-111.

Kuhn, H. 1953. "Extensive Games and the Problem of Information." Pp. 193-216 in H. Kuhn and A. Tucker (eds.), *Contributions to the Theory of Games*, Vol. II. Princeton, N.J.: Princeton University Press.

Migdalas, A., P.M. Pardalos, and P. Varbrand. 1998. *Multilevel Optimization: Algorithms and Applications*. Dordrecht, Germany: Kluwer.

Raiffa, H. 1968. *Decision Analysis*. Reading, Mass.: Addison-Wesley.

Salmerón, J., K. Wood, and R. Baldick. 2004. "Analysis of Electric Grid Security Under Terrorist Threat," *IEEE Transactions on Power Systems* 19(2):905-912.

Tintner, G. 1960. "A Note on Stochastic Linear Programming." *Econometrica* 28(2):490-495.

Ville, J. 1938. "Sur la theorie generale des jeux ou intervient l'habilite des joueurs." Pp. 105-113 in E. Borel et al. (eds.), *Traite du calcul des probabilites et de ses applications*, Vol. II. Paris: Gautheir-Villars.

von Neumann, J. 1928. "Zur Theorie der Gesellshaftsphiele." *Annals of Mathemetics* 100:295-320.

The White House. 2004. Homeland Security Presidential Directive 10 [HSPD-10]: *Biodefense for the 21st Century*. Available at www.fas.org/irp/offdocs/nspd/hspd-10.html. Accessed January 16, 2008.

The White House. 2007. Homeland Security Presidential Directive 18 [HSPD-18]: *Medical Countermeasures Against Weapons of Mass Destruction*. Available at www.fas.org/irp/offdocs/nspd/hspd-18.html. Accessed January 16, 2008.

Appendix F

Combining Game Theory and Risk Analysis in Counterterrorism: A Smallpox Example

David L. Banks
Professor, Institute of Statistics and Decision Sciences
Duke University, Durham, North Carolina

Steven Anderson
Director, Office of Biostatistics and Epidemiology
Center for Biologics Evaluation and Research
U.S. Food and Drug Administration, Rockville, Maryland

Abstract: Federal agencies have finite resources. Even for critical purposes related to counterterrorism, resources must be allocated in the most effective ways possible. Statistical risk analysis can help by accounting for uncertainties in the costs and benefits of particular efforts, and game theory can help by accounting for the fact that terrorists adapt their attacks in response to homeland defense initiatives. This paper describes a procedure that uses risk analysis to generate random payoff matrices for game theory solution, and then pools the solutions from multiple realizations of the payoff matrix to estimate the probability that a given play is optimal with respect to one of several criteria. The strategy is illustrated for risk management in the context of a simplified model of the threat of smallpox attack.

1. INTRODUCTION

The U.S. government wishes to invest its resources as wisely as possible in defense. Each wasted dollar diverts money that could be used to harden crucial vulnerabilities, prevents investment in future economic growth, and increases taxpayer burden. This is a classic conflict situation; a good strategy for the player with fewer resources is to leverage disproportionate resource investment by its wealthy opponent. That strategy rarely wins, but it makes the conflict sufficiently debilitating that the wealthy opponent may be forced to consider significant compromises.

Game theory is a traditional method for choosing resource investments in conflict situations. The standard approach requires strong assumptions about the availability of mutual information and the rationality of both opponents. Empirical research by many people (e.g., Kahneman and Tversky, 1972) shows that these assumptions fail in practice, leading to the development of modified theories with weaker assumptions or the use of prior probabilities in the spirit of Bayesian decision theory.

This paper considers both traditional game theory (minimax solution for a two-person zero-sum game in normal form) and also a minimum expected loss criterion appropriate for extensive-form games with prior probabilities. However, we emphasize that for terrorism, the zero-sum model is at best an approximation; the valuation of the wins and the losses is likely to differ between the opponents.

Game theory requires numerical measures of payoffs (or losses) that correspond to particular sets of decisions. In practice, those payoffs are rarely known. Statistical risk analysis allows experts to determine reasonable probability distributions for the random payoffs. This paper shows how risk analysis can support game theory solutions, and how Monte Carlo methods provide insight into the optimal game theory solutions in the presence of uncertainty about payoffs.

Our methodology is demonstrated in the context of risk management for a potential terrorist attack using the smallpox virus. The analysis we present here is a simplified version that aims at methodological explanation rather than analysis or justification of specific healthcare policies. As a tabletop exercise, the primary aim is only to provide a blueprint for a more rigorous statistical risk analysis. The underlying assumptions, modeling methods used here, and any results or discussion of the modeling are based on preliminary and unvalidated data and do not represent the opinion of the FDA, the Department of Health and Human Services or any branch of the U.S. government.

NOTE: Reprinted, with permission, from *Statistical Methods in Counterterrorism: Game Theory, Modeling, Syndromic Surveillance, and Biometric Authenticationon*. G. Wilson, and D. Olwell (eds.), Springer, 2006. pp. 9-22.

2. GAME THEORY FOR SMALLPOX

The smallpox debate in the United States has focused upon three kinds of attack and four kinds of defense. The three attack scenarios suppose that there might be:

- no smallpox attack
- a lone terrorist attack on a small area (similar to the likely scenario for the anthrax letters)
- a coordinated terrorist attack upon multiple population centers.

The four defense scenarios that have been publicly considered by United States agency officials are:

- stockpile smallpox vaccine
- stockpile vaccine and develop biosurveillance capabilities
- stockpile vaccine, develop biosurveillance, and inoculate key personnel
- provide mass vaccination to non-immunocompromised citizens in advance.

Although there are many refinements that can be considered for both the attack and the defense scenarios, these represent the possibilities discussed in the public meetings held in May and June 2002 (McKenna, 2002).

Suppose that analysts used game theory as one tool to evaluate potential defense strategies. Then the three kinds of attack and four kinds of defense determine a classic normal-form payoff matrix for the game [see Table 1].

The C_{ij} entries are the costs (or payoffs) associated with each combination of attack and defense, and we have used abbreviated row and column labels to identify the defenses and attacks, respectively, as described before.

For each of the 12 attack-defense combinations, there is an associated cost. These costs may include dollars, human lives, time, and other resources. For our calculation, all of these costs are monetized, according to principles detailed in Section 3. And the monetized value of a human life is set to $750,000, following the Department of Transportation's human capital model that estimates value from average lost productivity (non-market approaches tend to give larger values).

Note that there is very large uncertainty in the C_{ij} values. Portions of the cost (e.g., those associated with expenses already entailed) may be known, but the total cost in each cell is a random variable. These random variables are not independent, since components of the total cost are common to multiple cells. Thus it is appropriate to regard the entire game theory table as a multivariate random variable whose joint distribution is required for a satisfactory analysis that propagates uncertainty in the costs through to uncertainty about best play.

TABLE 1 Attack-Defense Cost Matrix

	No Attack	Single Attack	Multiple Attack
Stockpile Vaccine	C_{11}	C_{12}	C_{13}
Biosurveillance	C_{21}	C_{22}	C_{23}
Key Personnel	C_{31}	C_{32}	C_{33}
Everyone	C_{41}	C_{42}	C_{43}

Classical game theory (cf. Myerson 1991, Chapter 3) determines the optimal strategies for the antagonists via the minimax theorem. This theorem asserts that for any two-person cost matrix in a strictly competitive game (which is the situation for our example), there is an equilibrium strategy such that neither player can improve their expected payoff by adopting a different attack or defense. This equilibrium strategy may be a pure strategy, in which case optimal play is a specific attack-defense pair. This happens when the attack that maximizes the minimum damage and the defense that minimizes the maximum damage coincide in the same cell. Otherwise, the solution is a mixed strategy, in which case the antagonists pick attacks and defenses according to a probability distribution that must be calculated from the cost matrix. There may be multiple equilibria that achieve the same expected payoff, and for large matrices it can be difficult to solve the game.

Alternatively, one can use Bayesian decision theory to solve the game. Here a player puts a probability distribution over the actions of the opponent, and then chooses their own action so as to minimize the expected cost (cf. Myerson 1991, Chapter 2). Essentially, one just multiplies the cost in each row by the corresponding probability, sums these by row, and picks the defense with the smallest sum. This formulation is easier to solve, but it requires one to know or approximate the opponent's probability distribution and it does not take full account of the mutual strategic aspects of adversarial games (i.e., the assigned probabilities need not correspond to any kind of "if I do this then he'll do that" reasoning). Bayesian methods are often used in extensive-form games, where players make their choices over time, conditional on the actions of their opponent.

In developing our analysis of the smallpox example we make two assumptions about time. First, we use only the information available by June 1, 2002; subsequent information on the emerging program costs is not included. This keeps the analysis faithful in spirit to the decision problem actually faced by U.S. government policy makers in the spring of 2002 (their initial plan was universal vaccination, but ultimately they chose the third scenario with stockpiling, biosurveillance, and very limited vaccination of some first responders). Second, all of the estimated cost forecasts run to October 1, 2007. The likelihood of changing geopolitical circumstances makes it unrealistic to attempt cost estimates beyond that fiscal year.

3. RISK ANALYSIS FOR SMALLPOX

Statistical risk analysis is used to estimate the probability of undesirable situations and their associated costs. In the same way that it is used in engineering (e.g., for assessing nuclear reactor safety; cf. Speed, 1985) or the insurance industry (e.g., for estimating the financial costs associated with earthquakes in a specific area; cf. Brillinger, 1993), this paper uses risk analysis to estimate the costs associated with different kinds of smallpox attack/defense combinations.

Risk analysis involves careful discussions with domain experts and structured elicitation of their judgments about probabilities and costs. For smallpox planning, this requires input from physicians, public health experts, mathematical epidemiologists, economists, emergency response administrators, government accountants, and other kinds of experts. We have not conducted the in-depth elicitation from multiple experts in each area that is needed for a fully rigorous risk analysis; however, we have discussed the cost issues with representatives from each area, and we believe that the estimates in this section are sufficiently reasonable to illustrate, qualitatively, the case for combining statistical risk analysis with game theory for threat management in the context of terrorism.

Expert opinion was typically elicited in the following way. Each expert was given a written document with background on smallpox epidemiology and a short description of the attacks and defenses considered in this paper. The expert often had questions; these were discussed orally with one of the authors and, to the extent possible, resolved on the basis of the best available information. Then the expert was asked to provide a point estimate of the relevant cost or outcome and the range in which that value would be expected to fall in 95% of similar realizations of the future. If these values disagreed with those from other experts, then the expert was told of the discrepancy and invited to alter their opinion. Based on point estimate and the range, the authors and the expert chose a distribution function with those parameters which also respected real-world requirements for positivity, integer values, known skew, or other properties. As the last step in the interview, the expert was given access to all the other expert opinions obtained to that point and asked if there were any that seemed questionable; this led to in one case to an expert being recontacted and a subsequent revision of the elicitation. But it should be emphasized that these interviews were intended to be short, and did not use the full range of probes, challenges, and checks that are part of serious elicitation work.

The next three subsections describe the risk analysis assumptions used to develop the random costs for the first three cells (C_{11}, C_{21}, C_{31}) in the game theory payoff matrix. Details for developing the costs in the other cells are available from the authors. These assumptions are intended to be representative, realistic, and plausible, but additional input by experts could surely improve upon them. Many of the same costs arise in multiple cells, introducing statistical dependency among the entries. (That is, if a given random payoff matrix assumes an unusually large cost for stockpiling in one cell of the random table, then the same high value should appear in all other cells in which stockpiling occurs.)

3.1 Cell (1,1): Stockpile Vaccine/No Attack Scenario

Consider the problem of trying to estimate the costs associated with the (1,1) cell of the payoff matrix, which corresponds to no smallpox attack and the stockpiling of vaccine. This estimate involves combining costs with very different levels of uncertainty.

At the conceptual level, the cost C_{11} is the sum of four terms:

$$C_{11} = \mathrm{ET}_{\mathrm{dry}} + \mathrm{ET}_{\mathrm{Avent}} + \mathrm{ET}_{\mathrm{Acamb}} + \mathrm{VIG} + \mathrm{PHIS},$$

where $\mathrm{ET}_{\mathrm{dry}}$ and $\mathrm{ET}_{\mathrm{Avent}}$ are the costs of efficacy and safety testing for the Dryvax and Aventis vaccines, respectively; $\mathrm{ET}_{\mathrm{Acamb}}$ is the cost of new vaccine production and testing from Acambis; VIG is the cost of producing sufficient doses of vaccinia immune globulin to treat adverse reactions and possible exposures; and PHIS is the cost of establishing the public healthcare infrastructure needed to manage this stockpiling effort.

There is no uncertainty about $\mathrm{ET}_{\mathrm{Acamb}}$; the contract fixes this cost at \$512 million. But there is substantial uncertainty about $\mathrm{ET}_{\mathrm{dry}}$ and $\mathrm{ET}_{\mathrm{Avent}}$ since these entail clinical trials and may require follow-on studies; based on discussions with experts, we believe these costs may be realistically modeled as independent uniform random variables, each ranging between \$2 and \$5 million. There is also large uncertainty about the cost for producing and testing sufficient doses of VIG to be prepared for a smallpox attack; our discussions suggest this is qualitatively described by a normal random variable with mean \$100 million and a standard deviation of \$20 million. And there is great uncertainty about PHIS (which includes production of bifurcated inoculation needles, training, storage costs, shipment readiness costs, etc.); based on the five-year operating budget of other government offices with analogous missions, we assume this cost is normally distributed with mean \$940 million and standard deviation \$100 million.

3.2 Cell (2,1): Biosurveillance/No Attack Scenario

Biosurveillance programs are being piloted in several major metropolitan areas. These programs track data, on a daily basis, from emergency room admission records in order to quickly discover clusters of disease symptoms that suggest bioterrorist attack. Our cost estimates are based upon discussions with the scientists working in the Boston area (cf. Ross et al., 2002) and with the Pittsburgh team that developed monitoring procedures for the Salt Lake City Olympic games.

The cost C_{21} includes the cost C_{11} since this defense strategy uses both stockpiling of vaccine and increased biosurveillance. Thus

$$C_{21} = C_{11} + \text{PHIB} + \text{PHM} + \text{NFA} \cdot \text{FA}$$

where PHIB is the cost of the public health infrastructure needed for biosurveillance, including the data input requirements and software; PHM is the cost of a public health monitoring center, presumably at the Centers for Disease Control, that reviews the biosurveillance information on a daily basis; NFA is the number of false alarms from the biosurveillance system over five years of operation; and FA is the cost of a false alarm.

For this exercise, we assume that PHIB is normally distributed with mean $900 million and standard deviation $100 million (for a five-year funding horizon); this is exclusive of the storage, training, and other infrastructure costs in PHIS, and it includes the cost of hospital nursing-staff time to enter daily reports on emergency room patients with a range of disease symptoms (not just those related to smallpox). PHM is modeled as a normal random variable with mean $20 million and standard deviation $4 million (this standard deviation was proposed by a federal administrator, and may understate the real uncertainty).

False alarms are a major problem for monitoring systems; it is difficult to distinguish natural contagious processes from terrorist attacks. We expect about one false alarm per month over five years in a national system of adequate sensitivity, and thus FA is taken to be a Poisson random variable with mean 60. The cost for a single false alarm is modeled as a normal random variable with mean $500,000 and standard deviation $100,000.

3.3 Cell (3,1): Key Personnel/No Attack Scenario

One option, among several possible policies that have been discussed, is for the United States to inoculate about 500,000 key personnel, most of whom would be first-responders in major cities (i.e., emergency room staff, police, and public health investigators who would be used to trace people who have come in contact with carriers). If chosen, this number is sufficiently large that severe adverse reactions become a statistical certainty.

The cost of this scenario subsumes the cost C_{21} of the previous scenario, and thus

$$C_{31} = C_{21} + (\text{NKP} \times \text{IM}/25000) + (\text{PAE} \times \text{NKP} \times \text{AEC})$$

where NKP is the number of key personnel; IM is the cost of the time and resources needed to inoculate 25,000 key personnel and monitor them for adverse events; PAE is the probability of an adverse event; and AEC is the average cost of one adverse event. We assume that NKP is uniformly distributed between 400,000 and 600,000 (this reflects uncertainty about how many personnel would be designated as "key"). The IM is tied to units of 25,000 people, since this is a one-time cost and represents the number of people that a single nurse might reasonably inoculate and maintain records upon in a year. Using salary tables, we approximate this cost as a normal random variable with mean $60,000 and standard deviation $10,000.

The probability of an adverse event is taken from Anderson (2002), which is based upon Lane et al. (1970); the point estimate for all adverse events is .293, but since there is considerable variation and new vaccines are coming into production, we have been conservative about our uncertainty and assumed that the probability of an adverse event is uniformly distributed between .15 and .45. Of course, most of these events will be quite minor (such as local soreness) and would not entail any real economic costs.

The AEC is extremely difficult to estimate. For purposes of calculation, we have taken the value of a human life to be $2.86 million (the amount used by the National Highway Transportation Safety Administration in cost-benefit analyses of safety equipment). But most of the events involve no cost, or perhaps a missed day of work that has little measurable impact on productivity. After several calculations and consultations, this analysis assumes that *AEC* can be approximated as a gamma random variable with mean $40 and standard deviation $100 (this distribution has a long right tail).

4. ANALYSIS

The statistical risk analysis used in Section 3, albeit crude, shows how expert judgment can generate the random payoff matrices. The values in the cells of such tables are not independent, since many of the cost components are shared between cells. In fact, it is appropriate to view the table as a matrix-valued random variable with a complex joint distribution.

Random tables from this joint distribution can be generated by simulation. For each table, one can apply either the minimax criterion to determine an optimal strategy in the sense of von Neumann and Morgenstern (1944), or a minimum expected loss criterion to determine an optimal solution in the sense of Bayesian decision theory (cf. Myerson, 1991, Chapter 2). By doing this repeatedly, for many different random tables, one can estimate the proportion of time that each defense strategy is superior.

Additionally, it seems appropriate to track not just the number of times a defense strategy is optimal, but also weight this count by some measure of the difference between the costs of the game under competing defenses. For example, if two defenses yield game payoffs that differ only by an insignificant amount, it seems unrealistic to give no credit to the second-best strategy. For this reason we also use a scoring algorithm in which the score a strategy receives depends upon how well-separated it is from the optimal strategy.

Specifically, suppose that defense strategy i has value V_i on a given table. Then the score S_i that strategy i receives is

$$S_i = 1 - V_i/\{max\ V_j\}$$

and this ensures that strategies are weighted to reflect the magnitude of the monetized savings that accrue from using them. The final rating of the strategies is obtained by averaging their scores from many random tables.

4.1 Minimax Criterion

We performed the simulation experiment described above 100 times and compared the four defense strategies in terms of the minimax criterion. Although one could certainly do more runs, we believe that the approximations in the cost modeling are so uncertain that additional simulation would only generate spurious accuracy.

Among the 100 runs, we found that the Stockpile strategy won 9 times, the Biosurveillance strategy won 24 times, the Key Personnel strategy won 26 times, and the Vaccinate Everyone strategy won 41 times. This lack of a clear winner may be, at some intuitive level, the cause of the widely different views that have been expressed in the public debate on preparing for a smallpox attack.

If one uses scores, the results are even more ambiguous. The average score for the four defense strategies ranged between .191 and .326, indicating that the expected performances were, on average, quite similar.

From a public policy standpoint, this may be a fortunate result. It indicates that in terms of the minimax criterion, any decision is about equally defensible. This gives managers flexibility to incorporate their own judgment and to respond to extrascientific considerations.

4.2 Minimum Expected Loss Criterion

The minimax criterion may not be realistic for the game theory situation presented by the threat of smallpox. In particular, the normal-form game assumes that both players are ignorant of the decision made by their opponent until committed to a course of action. For the smallpox threat, there has been a vigorous public discussion on what preparations the United States should make. Terrorists know what the United States has decided to do, and presumably this will affect their choice of attack. Therefore the extensive-form version of game theory seems preferable. This form can be thought of as a decision tree, in which players alternate their moves. At each stage, the player can use probabilistic assessments about the likely future play of the opponent.

The minimum expected loss criterion requires more information that does the minimax criterion. The analyst needs to know the probabilities of a successful smallpox attack conditional on the U.S. selecting each of the four possible defenses. This is difficult to determine, but we illustrate how one can do a small sensitivity analysis that explores a range of probabilities for smallpox attack.

Table 2 shows a set of probabilities that we treat as the baseline case. We believe it accords with a prudently cautious estimate of the threat of a smallpox attack. To interpret Table 2, it says that if the United States were to only stockpile vaccine, then the probability of no smallpox attack is .95, the probability of a single attack is .04, and the probability of multiple attacks is .01. Similarly, one reads the attack probabilities for other defenses across the row. All rows must sum to one.

The minimum expected loss criterion multiplies the probabilities in each row of Table 2 by the corresponding costs in the same row of Table 1, and then sums across the columns. The criterion selects the defense that has the smallest sum.

As with the minimax criterion, one can simulate many payoff tables and then apply the minimum expected loss criterion to each. In 100 repetitions, Stockpile won 96 times, Biosurveillance won 2 times, and Vaccinate Everyone won twice. The scores showed roughly the same pattern, strongly favoring the Stockpile defense.

We now consider two alternative sets of probabilities, shown in Table 3 and Table 4. Table 3 is more pessimistic, and has larger attack probabilities. Table 4 is more optimistic, and has smaller attack probabilities. A serious sensitivity analysis would investigate many more tables, but our purpose is illustration and we doubt that the quality of the assessments that underlie the cost matrix can warrant further detail.

For Table 3, 100 simulation runs found that Stockpile won 15 times, Biosurveillance won 29 times, Key Personnel won 40 times, and Vaccinate Everyone won 16 times. In contrast, for Table 4, the Stockpile strategy won 100 times in 100 runs.

The scores for Table 3 ranged from 18.2 to 38.8, which are quite similar. In contrast, for Table 4 nearly all the weight of the score was on the Stockpile defense. These results show that the optimal strategy is sensitive to the choice of probabilities used in the analysis. Determining those prob-

TABLE 2 Baseline Probabilities of Attack Given Different Defenses

	No Attack	Single Attack	Multiple Attack
Stockpile Vaccine	0.95	0.04	0.01
Biosurveillance	0.96	0.035	0.005
Key Personnel	0.96	0.039	0.001
Everyone	0.99	0.005	0.005

TABLE 3 Pessimistic Probabilities of Attack Given Different Defenses

	No Attack	Single Attack	Multiple Attack
Stockpile Vaccine	0.70	0.20	0.10
Biosurveillance	0.80	0.15	0.05
Key Personnel	0.85	0.10	0.05
Everyone	0.90	0.05	0.05

TABLE 4 Optimistic Probabilities of Attack Given Different Defenses

	No Attack	Single Attack	Multiple Attack
Stockpile Vaccine	0.98	0.01	0.01
Biosurveillance	0.99	0.005	0.005
Key Personnel	0.99	0.005	0.005
Everyone	0.999	0.0005	0.0005

abilities requires input from the intelligence community and the judgment of senior policy-makers.

5. CONCLUSIONS

This paper has outlined an approach combining statistical risk analysis with game theory in order to evaluate defense strategies that have been considered for the threat of smallpox. We believe that this approach may offer a useful way of structuring generic problems in resource investment for counterterrorism.

The analysis in this paper is incomplete:

1. We have focused upon smallpox, because the problem has been framed rather narrowly and quite definitively by public discussion. But a proper game theory analysis would not artificially restrict the options of the terrorists, and should consider other attacks, such as truck bombs, chemical weapons, other diseases, and so forth (which would get difficult, but there may be ways to approximate). It can be completely misleading to seek a local solution, as we have done.
2. Similarly, we have not fully treated the options of the defenders. For example, heavy investment in intelligence sources is a strategy that protects against many different kinds of attacks, and might well be the superior solution in a less local formulation of the problem.
3. We have not considered constraints on the resources of the terrorists. The terrorists have limited resources and can invest in a portfolio of different kinds of attacks. Symmetrically, the U.S. can invest in a portfolio of defenses. This aspect of the problem is not addressed—we assume that both parties can fund any of the choices without sacrificing other goals.
4. The risk analysis presented here, as discussed previously, is not adequate to support public policy formulation.

Nonetheless, despite these limitations, the methodology has attractive features. First, it is easy to improve the quality of the result through better risk analysis. Second, it automatically raises issues that have regularly emerged in policy discussions. And third, it captures facets of the problem that are not amenable to either game theory or risk analysis on their own, because classical risk analysis is not used in adversarial situations and because classical game theory does not use random costs.

NOTES: BACKGROUND ON SMALLPOX

Although the probability that the smallpox virus (*Variola major*) might be used against the U.S. is thought to be small, the public health and economic impact of even a limited release would be tremendous. Any serious attack would probably force mass vaccination programs, causing additional loss of life due to adverse reactions. Other economic consequences could easily be comparable to those of the attacks of September 11, 2001.

A smallpox attack could potentially be initiated through infected humans or through an aerosol (Henderson et al., 1999). In 12-14 days after natural exposure patients experience fever, malaise, body aches, and a body rash (Fenner et al., 1988). During the symptomatic stages of the disease the patient can have vesicles in the mouth, throat, and nose that rupture to spread the virus during a cough or sneeze.

Person-to-person spread usually occurs through inhalation of virus-containing droplets or from close contact with an infected person. As the disease progresses the rash spreads to the head and extremities and evolves into painful, scarring vesicles and pustules. Smallpox has a mortality rate of approximately 30%, based on data from the 1960s and 1970s (Henderson, 1999).

Various mathematical models of smallpox spread exist and have been used to forecast the number of people infected under different exposure conditions and different public health responses (cf. Kaplan, Craft, and Wein, 2002; Meltzer et al., 2001). There is considerable variation in the predictions from these models, partly because of differing assumptions about the success of the "ring vaccination" strategy that has been planned by the Centers for Disease Control (2002), and this is reflected in the public debate on the value of preemptive inoculation versus wait-and-see preparation. However, the models are in essential agreement that a major determinant of the size of the epidemic is the number of people who are exposed in the first attack or attacks.

The current vaccine consists of live vaccinia or cowpox virus and is effective at preventing the disease. Also, vaccination can be performed within the first 2 to 4 days post exposure to reduce the severity or prevent the occurrence of the disease (Henderson, 1999).

But vaccination is not without risk; the major complications are serious infections and skin disease such as progressive vaccinia, eczema vaccinatum, generalized vaccinia, and encephalitis. Approximately 12 people per million have severe adverse reactions that require extensive hospitalization, and about one-third of these die—vaccinia immune globulin (VIG) is the recommended therapy for all of these reactions except encephalitis. Using data from Lane et al. (1970), we estimate that 1 in 71,429 people suffer postvaccinial encephalitis, 1 in 588,235 suffer progressive vaccinia, 1 in 22,727 suffer eczema vaccinatum, and 1 in 3,623 suffer generalized vaccinia. Additionally, 1 in 1,656 people suffer accidental infection (usually to the eye) and 1 in 3,289 suffer some other kind of mild adverse event, typically requiring a person to miss a few days of work. (Other studies give somewhat different numbers; cf. Neff et al., 1967a, 1967b). People who have previously been successfully vaccinated for smallpox are less likely to have adverse reactions, and people who are immunocompromised (e.g., transplant patients, those with AIDS) are at greater risk for adverse reactions (cf. Centers for Disease Control, 2002, Guide B, parts 3, 5, and 6).

Because the risk of smallpox waned in the 1960s, vaccination of the U.S. population was discontinued in 1972. It is believed that the effectiveness of a smallpox vaccination diminishes after about 7 years, but residual resistance persists even decades later. It has been suggested that people who were vaccinated before 1972 may be substantially protected against death, if not strongly protected against contracting the disease (cf. Cohen, 2001).

The U.S. currently has about 15 million doses of the Wyeth Dryvax smallpox vaccine available. The vaccine was made by scarification of calves with the New York City Board of Health strain and fluid containing the vaccinia virus was harvested by scraping (Rosenthal et al., 2001). Recent clinical trials on the efficacy of diluted vaccine indicate that both the five-fold and ten-fold dilutions of Dryvax achieve a take rate (i.e., a blister forms at the inoculation site, which is believed to be a reliable indicator of immunization) of at least 95%, so the available vaccine could be administered to as many as 150 million people should the need arise (cf. Frey et al., 2002; NIAID, 2002).

The disclosure by the pharmaceutical company Aventis (Enserink, 2002) of the existence in storage of 80 to 90 million doses of smallpox vaccine that were produced more than 30 years ago has added to the current stockpile. Testing is being done on the efficacy of the Aventis vaccine stock, including whether it, too, could be diluted if needed.

Contracts to make new batches of smallpox vaccine using cell culture techniques have been awarded to Acambis. The CDC amended a previous contract with Acambis in September 2001 to ensure production of 54 million doses by late 2002. Another contract for the production of an additional 155 million doses was awarded to Acambis in late November 2001, and the total cost of these contracts is $512 million. After production, additional time may be needed to further test the safety and efficacy of the new vaccine (cf. Rosenthal et al., 2001).

REFERENCES

Anderson, S. (2002). "A risk-benefit assessment of smallpox and smallpox vaccination," Technical Report, Office of Biostatistics and Epidemiology, Center for Biologics Evaluation and Research, U.S. Food and Drug Administration, Rockville, MD, 2002.

Brillinger, D.R. (1993). "Earthquake risk and insurance," *EnviroMetrics*, 4, 1-21.

Centers for Disease Control, (2002). *Smallpox response plan and guidelines (Version 3.0)*, www.bt.cdc.gov/agent/smallpox/response-plan/index.asp.

Cohen, J. (2001). "Smallpox vaccinations: How much protection remains?" *Science*, 294, 985.

Enserink, M. (2002). "New cache eases shortage worries," *Science*, 296 (April 5) 25-26.

Fenner, F., Henderson, D.A., Arita, I., Jezek, Z., Ladnyi, I.D. (1988). *Smallpox and Its Eradication*, World Health Organization, Geneva.

Frey, S.E., Couch, R.B, Tacket, C.O., Treanor, J.J., Wolff, M., Newman, F.K. Atmar, R.L., Edelman, R., Nolan, C.M., Belshe, R.B. (2002). "Clinical responses to undiluted and diluted smallpox vaccine," *New England Journal of Medicine*, 346:17, 1265-1274.

Halloran, M.E., Haber, M., Longini, I.M, Jr., and Struchiner, C.J. (1991). "Direct and indirect effects in vaccine efficacy and effectiveness," *American Journal of Epidemiology*, 133, 323-331.

Henderson, D.A. (1999). "Smallpox: Clinical and epidemiological features," *Emerging Infectious Diseases*, 5, 537-539.

Henderson, D.A., Ingelsby, T.V., Bartlett, J.G., Ascher, M.S., Eitzen, E., Jahrling, P.B., Hauer, J., Layton, M., McDade, J., Osterholm, M.T., O'Toole, T., Parker, G., Perl, T., Russel, P.K., and Tonat, K. (1999). "Smallpox as a biological weapon—Medical and public health management," *Journal of the American Medical Association*, 281:22, 2127-2137.

Kahnemann, D., and Tversky, A. (1972). "Subjective probability: A judgment of representativeness," *Cognitive Psychology*, 3, 430-454.

Kaplan, E., Craft, D.L., and Wein, W.M. (2002). "Emergency response to a smallpox attack: The case for mass vaccination," *Proceedings of the National Academy of Sciences*, 99, 5237-5240.

Lane, M.J., Ruben, F.L., Neff, J.M., and Millar, J.D. (1970). "Complications of smallpox vaccination, 1968: Results of ten statewide surveys," *Journal of Infectious Diseases*, 122, 303-309.

McKenna, M.A.J. (2002). "No mass smallpox vaccinations, panel recommends," *Atlanta Journal-Constitution*, June 21, p. 1.

Meltzer, M.I., Damon, I., LeDuc, J.W., and Millar, J.D. (2001). "Modeling potential responses to smallpox as a bioterrorist weapon," *Emerging Infectious Diseases*, 7, 201-208.

Myerson, R.B. (1991). *Game Theory: Analysis of Conflict*, Harvard University Press, Cambridge, MA, 1991.

NIAID. (2002). "NIAID study results support diluting smallpox vaccine stockpile to stretch supply," *NIAID News*, March 28. National Institute of Allergy and Infectious Diseases, www.niaid.nih.gov/newsroom/releases/smallpox.htm.

Neff, J.M., Lane, J., Pert, J.P., et al. (1967). "Complications of smallpox vaccination, I: National survey in the United States, 1963," *New England Journal of Medicine*, 276, 1-8.

Neff, J.M., Levine, R.H., Lane, J.M., et al. (1967). "Complications of smallpox vaccination, United States, 1963, II: Results obtained from four statewide surveys," *Pediatrics*, 39, 16-923.

Rosenthal, S.R., Merchilinsky, M., Kleppinger, C., and Goldenthal, K.L. (2001). "Developing new smallpox vaccines," *Emerging Infectious Diseases*, 7, 920-926.

Ross, L., Kleinman, K., Dashevsky, I., Adams, C., Kludt, P., DeMaria, A., Jr., and Platt, R. (2002). "Use of automated ambulatory-care encounter records for detection of acute illness clusters, including potential bioterrorism events," *Emerging Infectious Diseases*, 8, 753-760.

Speed, T.P. (1985). "Probabilistic risk assessment in the nuclear industry: WASH-1400 and Beyond," in *Proceedings of the Berkeley Conference in Honor of Jerzy Neyman and Jack Kiefer*, Vol. 2, L. LeCam and R. Olshen, eds., Wadsworth, Pacific Grove, CA, pp.~173-200.

Treaster, J.B., (2002). "The race to predict terror's costs," *New York Times*, Sept. 1, section 3, p. 1.

von Neumann, J., and Morgenstern, O. (1944). *Theory of Games and Economic Behavior*, Princeton University Press, Princeton NJ.

Appendix G

On the Quantification of Uncertainty and Enhancing Probabilistic Risk Analysis

Nozer D. Singpurwalla
Professor, Department of Statistics
George Washington University, Washington, D.C.

PREAMBLE

This appendix consists of two parts. In Part 1, we overview some commonly used approaches for quantifying uncertainty. The overview is necessarily terse, but adequate references are provided. Herein we introduce the notions of chance, probability, likelihood, belief, and plausibility, terms that commonly arise in the context of risk analysis. Also mentioned here are the notions of consequences and utilities, both of which are germane to risk analysis and risk management. Part 1 can serve as a supplement to the "Lexicon of Probabilistic Risk Assessment Terms" given in Appendix A of this report.

In Part 2 we put forth some thoughts and ideas for enhancing PRA (Probabilistic Risk Analysis) with some statistical and decision theoretic methodologies that are available in the literature, and which could be advantageously invoked. We close this section by alluding to the possibility of some new research in PRA, namely, the development of an architecture for adversarial *risk analysis* and *decision making in vague* (or fuzzy) *environments*.

It is our hope that this appendix will fill in any gaps of interpretation of the Lexicon that is given in the text, so that this appendix and the Lexicon of Appendix A are linked. To better facilitate a broad based appreciation of the material presented here, this appendix has been deliberately cast in a conversational style. That is, mathematical notation has been avoided.

PART 1. APPROACHES TO QUANTIFYING UNCERTAINTY

Introduction

From a layperson's point of view, the term "risk" connotes the possibility that an undesirable event will occur. However, the modern technical meaning of the term is different. Here, *risk* is the sum of the product of one's personal *probabilities* (or the objective *chances*) of all possible outcomes (also known as *consequences*) of an action, and the *utilities* of each outcome. Probabilities and chances are ways to quantify uncertainty (i.e., the possibility mentioned above), and quantification is a necessary step for invoking the logical argument. Utilities are numerical values of the consequences of each outcome, on a zero to one scale. Indeed, utilities are probabilities and must therefore obey the rules (or the calculus) of probability (cf. Lindley, 1985, p. 56). They quantify one's preferences between consequences. Thus the modern notion of risk entails the twin notions of probability (or chance) and utility. Its computation via the sum of products rule mentioned above (cf. Morgeson et al. [2006] for a detailed application of this principle to terrorist risk assessment) is a consequence of the calculus of probability. The quantification of uncertainty by probability is, according to de Finetti (1972) and Lindley (1982), the only satisfactory way. Alternatives to probability, like Zadeh's (1979) *possibility*, do not lead to a prescription for the quantification of risk; this is one of its biggest drawbacks.

Chance and Probability: Metrics for Quantifying Uncertainty

The use of probability as a metric for quantifying uncertainty dates back to the 16th century. However, discussions about its meaning and interpretation continue until today. The distinction between chance and probability (cf. Good, 1990) is a consequence of such debates and discussions. In his review article, Kolmogorov (1969) wholeheartedly subscribes to probability as an objective *chance* that is agreed upon by all even though it can never be observed. It is defined as the limit of a relative frequency; the operational word being "limit." To Kolmogorov, chance and probability were synonymous, and thus the word chance does not appear in his writings. To de Finetti (1976) and others, like Savage (1972), probability is subjective and personal, and encapsulates ones disposition to a two-sided bet. De Finetti (1972) goes further by connecting chance and probability via his theorem on *exchangeable sequences* with the thesis that probability is to be seen as a two-sided bet about the unknown chance. The algebra (or the calculus) of probability is subscribed to by all (save the axiom of *countable additivity* which to de Finetti is unnecessary). Whereas an unobservable chance can be *estimated* via observed data (if available), probability can be made operational by monitoring one's disposition to a series of bets. One needs to monitor a series of bets to ensure that the bettor adheres to the calculus of probability; i.e. the bettor needs to be *coherent*.

Likelihood: A Weighting Function

The term *likelihood* has often been used as a substitute for chance and probability. However, the technical meaning of the term is different. Indeed, it can be seen that a likelihood is *not* a probability (or chance), and that a likelihood does not obey the calculus of probability. The notion of a likelihood arises in the context of making assessments of uncertainty in the light of new evidence (or data) using Bayes' Law. The likelihood is simply a weighting function that can be assigned either subjectively or via a probability model. The matter is subtle and warrants a detailed discussion that cannot be given here. We refer the reader to Singpurwalla (2006), Section 2.4.3, or to Singpurwalla (2007) for a more complete picture. The essence of this sub-section is that like chance and probability, the likelihood is, from a technical point of view, a distinct construct. Thus, caution should be used when it is used with the first two.

Probabilistic Risk Analysis

Probabilistic risk analysis—henceforth PRA—is a systematic way to assess and to invoke the calculus of probability. Its origins can be traced to the work done at Bell Telephone Laboratories on the launching of missiles (cf. Watson, 1961), and to the work done at the Boeing Scientific Laboratories on assessing the reliability of airplanes (cf. Hassl, 1965). The prominence of PRA grew with the dawning of the nuclear reactor era when it became the dominant tool for assessing the safety of nuclear reactors (cf. Barlow, et al., 1975). The driving tools behind a PRA are the *event trees* and *fault trees*, which are a graphical portrayal of the causes that lead up (or down) to an event of interest. At the terminus of such trees are the causes that trigger the event of interest; such causes are called the *basic events* of the trees. PRA is attractive to engineers and other scientists because of their inherent graphic feature, just as *Bayesian Belief Nets* (BBNs) are attractive to computer scientists. When all is said and done, both the PRA and the BBN are simply tools for assessing probabilities, and invoking the calculus of probability. They are devices for good book-keeping practices in probability calculations.

The distinction between chance and probability is germane to PRA, because each leads to a different paradigm for assessing risk. The former leads to the frequentist (or sample-theoretic) approach, the latter to the subjectivistic Bayesian approach. Under the frequentist approach, PRA can only be done when hard data on the basic events are at hand, and preferably a substantial amount. Such data could be easy to come by when one deals with conceptually repeatable events like failures in a population of items such as valves, electronics, and other such small gadgets. PRA under frequentist paradigm is most suitable for engineered systems like airplanes, automobiles, tanks, and nuclear reactors. By contrast, under the Bayesian approach to PRA, probabilities of the basic events need to be subjectively obtained via the elicitation, codification, modulation, and the fusion of expert testimonies (see, for example, Singpurwalla, 2006, Chapter 5). Because terrorist risk related events are not considered to be repeatable (to constitute an *ensemble*), PRA under the subjectivistic Bayesian paradigm appears to be relevant, not only in the contexts of biological agent risk analysis and other modes of terrorist risk (cf. Morgeson et al., 2006), but also for human health risk assessment from environmental hazards (cf. Nayak and Kundu, 2001, who also allude to a distinction between chance and probability vis-à-vis "variability" and "uncertainty").

The Dynamic Nature of Subjective Probability

With the above in place, some caveats about the subjective probabilities and their assessments need to be stated. Unlike chance—an objective entity—that is fixed for all time and agreed upon by all, subjective probability is personal to an individual (or a group acting as one), and can change from person to person. More important, it can change over time even for the same person. In other words, *subjective probability is dynamic*. It is assessed at some fixed point in time and the assessment is presumably based on the information at hand at that fixed point in time. As time marches on, new information could become available, and with it a possible change of probability. The position that subjective probability can be dynamic takes a more dramatic stand via the claim that it is not merely the availability of new information over time that brings about a change in probability. A change in probability could also be the result of a change in the psychological disposition of the individual whose betting behavior is assessed (cf. Ramsey, 1926). It is because of the above caveats that de Finetti (1974) in the introduction of his famous two-volume book on probability declares that: "Probability Does not Exist."

Alternatives to Chance and Probability

One, among the several, of Kolmogorov's (1933) notable achievements was that he freed probability from the debates and discussions of interpretation. He did this by axiomatizing probability. (The call to axiomatize probability can be traced to the German mathematician David Hilbert, Kolmogorov's dissertation supervisor, and to Sergei N. Bernstein). However, in order to axiomatize probability, Kolmogorov had to introduce an architecture, and it is aspects of this architecture that have paved the way for an entrance of alternatives to probability.

The mathematical architecture upon which the axiomatization of probability rests consists of a *sample space* (i.e., the set of all possible outcomes of a random phenomenon), and a *many to one mapping* (or a function) from the sample space to the real line. The mapping is known as a *random variable*. Probability is another mapping defined on the subsets of the sample space. It takes values between 0 and 1, and it abides by the addition and multiplication rules of probability. Kolmogorov's architecture subscribes to the *law of the excluded middle*. The essence of this law is that every element of the sample space can either belong, or not belong, to a particular sub-set of the sample space. In other words, any element of the sample space cannot simultaneously belong and not belong to any sub-set of the sample space. This happens when the sub-sets are sharp; that is, their boundaries are well defined.

Objections to Kolmogorov's architecture stem from two directions. The first is that in practice, especially when it comes to linguistic information, the law of the excluded middle turns out to be a restriction. In other words, requiring that sub-sets of the sample space have sharp boundaries is restrictive. One needs to entertain the possibility that the boundary of the said sub-sets could be vague or *fuzzy*. The second objection pertains to the requirement that the mapping from the sample space to the real line may be many to one. In practice, scenarios can arise wherein the said mapping needs to be *one to many*. Such scenarios can generally arise in the context of forensics, accident investigation, or failure diagnosis.

The need to entertain fuzzy sets has led Zadeh (1979) to propose an alternative to probability, namely, *possibility theory*. The calculus of possibility theory is different from that of probability theory; it parallels that of operations with fuzzy sets. Thus fuzzy set theory and possibility theory are often mentioned in the same vein. Regrettably, and despite Zadeh's persistent efforts, there has been no justification of the calculus of possibility theory. By contrast, the axioms of probability theory—the Kolmogorov axioms—have a foundation that is rooted in behavioristic phenomena. As a consequence, possibility theory has failed to provide a prescription for calculating risk. More important, it has been recently argued (cf. Singpurwalla and Booker, 2004) that it is possible to endow fuzzy sets with probability measures. This has made the role of possibility theory unnecessary.

The need to entertain scenarios involving one-to-many mappings has motivated Dempster (1968) to propose a generalization of probability measures, which he calls *belief* and *plausibility*; some details about these can be had from Singpurwalla and Wilson (2007) and the references therein. The net effect of these measures is that probability, instead of being a single number, is bounded above and below by what are known as *upper* and *lower probabilities* (also see Walley, 1991). A proposal for decision making based on upper and lower probabilities has been made by Giron and Rios (1980). Whereas this proposal lacks the force of coherence that decision making based on probabilities has, it may serve as a basis for risk analysis based on belief and plausibility. This possibility remains to be explored.

PART 2. ENHANCING PRA WITH BEST PRACTICES

The material of this part is linked with that of Part 1 wherein it was stated that probability and utility are two components of risk analysis, and that PRA was a tool to facilitate the assessment of probabilities of certain events, using the calculus of probability. A prescription for computing risk was also given, and it was stated that in the context of biological agent risk analysis PRA under the subjectivistic Bayesian paradigm would be the desired approach. The dynamic nature of subjective probability was mentioned and the need to ensure coherence of elicited probabilities was emphasized. The prescription for calculating risk as the sum of the product of probabilities and utilities was a consequence of the calculus of probability, and the fact that utilities are probabilities.

In the context of managing risk, one chooses that action for which the calculated risk is a minimum. This prescription for taking actions constitutes the basis of decision making under uncertainty (cf. Raiffa and Schlaifer, 1961) wherein *decision trees* play a role analogous to that of fault and event trees. That is, decision trees facilitate good book-keeping in the context of making decisions. Decision theorists are attracted to decision trees for the same reason that engineers liking fault trees, event trees, and PRA; graphics is the virtue of both. The important point to note is that generally, decision trees pertain to the flow of actions and events that are of relevance to a *single* decision maker. With the above as a perspective, the following enhancements to the current methods of using PRA for risk analysis and management can be suggested.

1. The elicited subjective probabilities should be tested to *ensure coherence* via more than a single query of the "expert."
2. The assessed subjective probabilities should be *modulated* to make adjustments for any inherent biases that the experts may have.
3. When the assessed subjective probabilities entail more than one expert—and this on principle should always be attempted—the expert testimonies should be *fused* in a manner that accounts for the correlations (positive or negative) among the experts.

Steps 2 and 3 above should be done formally via the calculus of probability. Details about how this can be done are given in Singpurwalla (2006, Chapter 5), wherein references to the original sources can be found. Some researchers (Cooke, 1991) argue strongly in favor of calibrating probabilities against empirical data as an alternative to modulation. The author disagrees that proper Bayesian methods for modulating assessed probabilities are not available. Philosophical issues aside, the calibration method suggested by Cooke requires empirical data; and in the absence of such data, modulating the assessed probabilities based on one's assessment of the expertise of the experts is a desirable option.

4. To many, a routine use of subjective probabilities and their accompanying paraphernalia of Bayesian methods in the context of PRA are objectionable; see, for example, Nayak and Kundu (2001). This is particularly acute when it comes to matters of public policy wherein some sense of objectivity becomes paramount. Thus whenever hard data on the basic events are available, *frequentist methods* should also be used, for no other reason than as a means of *calibrating* the Bayesian results.
5. Risk calculations based on subjective probabilities and Bayesian methods should be investigated for their *robustness* and *sensitivity* against the priors and the coding, modulating, and fusing mechanisms.
6. Much of the current work in PRA uses stylized metrics such as dollars or lives lost, for utilities. Statisticians routinely use squared error or the absolute error as the metrics of utility. Such metrics, while easy to implement, may not reflect the true preferences of a decision maker. Thus formal methods of *utility elicitation* as prescribed in the von Neumann and Morgernstern (1944) interpretation of utility should be considered. Endowing a PRA with utilities that are formally elicited will be a major step forward. This seems to be lacking.
7. In the context of terrorist risk assessment, be it biological or otherwise, the *layered defense and attack* concepts used in military science could be valuable; an inkling of these appears in Morgeson et al. (2006). Under a layered defense, the probability of penetration goes down with the number of layers, resulting in lower probability of a successful attack on an asset. The effect of all this would be an expansion of the event and fault trees and the assessment of several conditional probabilities.
8. Even though alternatives to probability have often been mentioned in the context of a PRA, there do not seem to be at hand concrete examples and illustrations demonstrating the viability of such alternatives. A possible reason behind this state of affairs could be the lack of awareness about the availability of some tools that are able to deal with decision making in a fuzzy environment, and in the presence of a one-to-many map. Singpurwalla and Booker (2004) and Giron and Rios (1980) allude to such tools. These tools, albeit unproven, offer a pathway toward enhancing the current PRA technology, and are worth attempting given the repeated calls for PRA under alternatives to probability.

It was mentioned before that the traditional decision trees which provide a prescription for action to mitigate the possibility of an adverse outcome were pertinent to a single decision maker. More important, the decision maker's opponent is considered to be nature, a benevolent adversary. The same is also true of fault trees and event trees, the staple tools of a PRA. Game theory comes into play when the adversary is not benevolent, like a terrorist. When such is the case the static decision, fault, and event trees need to be enhanced to incorporate adversarial behavior. Thus the graphics and the underlying mathematics of a PRA need to be modified so that they encapsulate adversarial actions. However doing so under

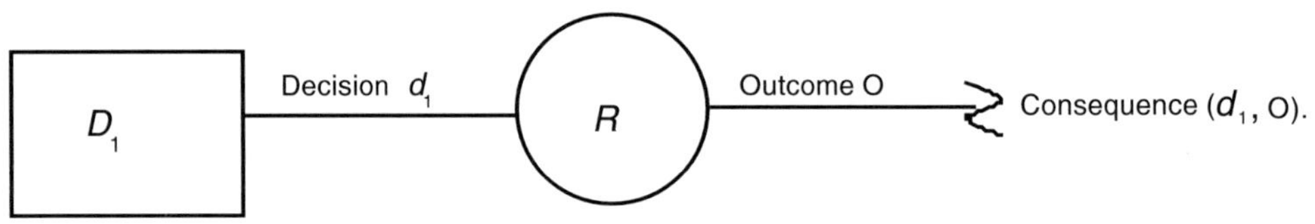

FIGURE G.1 Non-adversarial decision tree of D_1.

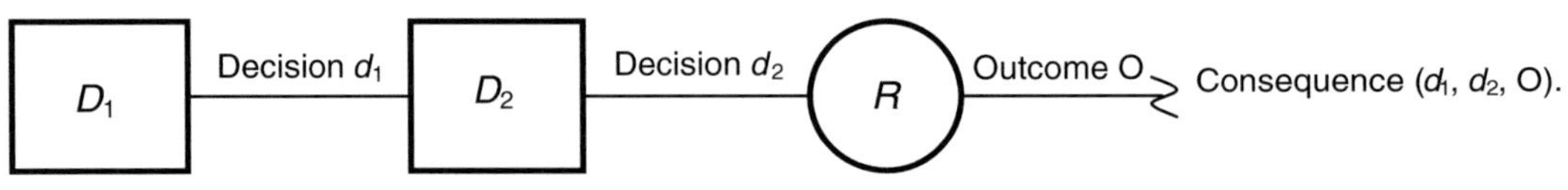

FIGURE G.2 Adversarial decision tree of D_1.

the umbrella of standard game theory would be problematic because of the matter of *infinite regress* (see for example, von Neumann and Morgenstern, 1944). A possible compromise would be to consider the use of an adversarial decision tree. An *adversarial decision tree* (cf. Lindley and Singpurwalla, 1991, 1993) portrays the schemata of adversarial decision making when the actions of each adversary are sequential. The layered attack and defense scenario mentioned above would serve as a suitable model that calls for an adversarial event, fault, and decision tree. Since the adversarial actions change over time, the underlying probabilities will need to be reassessed over time, and the dynamic nature of subjective probability allows for this constant reassessment.

To get some sense of what an adversarial decision tree would look like, consider Figures G.1 and G.2. The former has a single decision node, D_1, wherein D_1 encapsulates the actions of D_1, a single decision maker. Figure G.1 portrays the scenario of non adversarial decision making. By contrast, Figure G.2 which consists of two decision nodes D_1 and D_2, portrays the contemplated sequential actions of two decision makers, D_1, and his/her adversary D_2. The latter will supposedly (to D_1) act in the light of the actions of D_1 and their possible consequences. However, the decision tree itself pertains to the actions that D_1 should take, taking into consideration the possible actions of D_2. The overall aim is for D_1 to maximize his/her expected utility. Figure G.2 can be extended to cover the repeated actions of D_1 and D_2 over several cycles. However, the total number of cycles must be finite, or else the matter of infinite regress will begin to creep back. The decision nodes D_i, the decisions d_i, $i = 1, 2$, and the random node R of Figures G.1 and G.2 are conventional (see, for example Raiffa and Schlaifer, 1961).

REFERENCES

Barlow, R.E., H.B. Fussell, and N.D. Singpurwalla (eds.). 1975. *Reliability and Fault Tree Analysis: Theoretical and Applied Aspects of System Reliability and Safety Assessment.* Philadelphia, Pa.: SIAM.

Cooke, R.M. 1991. *Experts in Uncertainty: Opinion and Subjective Probability in Science.* New York: Oxford University Press.

de Finetti, B. 1972. *Probability, Induction and Statistics.* New York: Wiley.

de Finetti, B. 1974. *Theory of Probability.* New York: Wiley.

de Finetti, B. 1976. "Probability: Beware of Falsification!" *Scientia* 111:283-303.

Dempster, A.P. 1968. "A Generalization of Bayesian Inference." *Journal of the Royal Statistical Society*, Series B 30:205-247.

Giron, F., and S. Rios. 1980. "Quasi-Bayesian Behavior: A More Realistic Approach to Decision Making?" In J.B. Bernardo, M.H. De Groot, D.V. Lindley, and A.F.M. Smith (eds.), *Bayesian Statistics.* Valencia U.P.: University of Valencia Press.

Good, I.J. 1990. "Subjective Probability." In J. Eatwell, M. Milgate, and P. Newman (eds.), *The New Palgrave: Utility and Probability.* New York: Norton.

Hassl, D. 1965. "Advanced Concepts in Fault Tree Analysis." Paper presented at System Safety Symposium, sponsored by University of Washington and Boeing Company, Seattle, Washington.

Kolmogorov, A.N. 1933. *Foundations of the Theory of Probability.* New York: Chelsea Publishing.

Kolmogorov, A.N. 1969. "The Theory of Probability." Pp. 229-264 in A.D. Aleksandrov, A.N. Kolmogorov, M.A. Lavrentev (eds.), *Mathematics, Its Contents, Methods and Meaning*, Vol. 2, Part 3. Cambridge, Mass.: MIT Press.

Lindley, D.V. 1982. "Scoring Rules and the Inevitability of Probability." *International Statistical Review* 50(1):1-26.

Lindley, D.V. 1985. *Making Decisions*, 2nd ed. New York: Wiley.

Lindley, D.V., and N.D. Singpurwalla. 1991. "On the Evidence Needed to Reach Agreed Action Between Adversaries, with Application to Acceptance Sampling." *Journal of the Royal Statistical Association* 86(416):933-937.

Lindley, D.V., and N.D. Singpurwalla. 1993. "Adversarial Life Testing." *Journal of the Royal Statistical Society*, Series B 55(4):837-847.

Morgeson, J.D., V.A. Utgoff, M.A. Fainberg, and M. Keleher. 2006. "National Risk Assessment Pilot Project." Institute for Defense Analyses, Document D-3309. Arlington, Va.

Nayak, T.K., and S. Kundu. 2001. "Calculating and Describing the Uncertainty in Risk Assessment: The Bayesian Approach." *Human and Ecological Risk Assessment* 7(2):307-328.

Raiffa, H., and R. Schlaifer. 1961. *Applied Statistical Decision Theory.* Boston: Harvard University, Graduate School of Business Administration, Division of Research.

Ramsey, F.P. 1926. "Truth and Probability." In *Foundations of Mathematics and Other Logical Essays.* New York: Humanities Press.

Savage, L.J. 1972. *The Foundations of Statistics*, 2nd ed. New York: Dover.

Singpurwalla, N.D. 2006. *Reliability and Risk: A Bayesian Perspective.* Hoboken, N.J.: Wiley.

Singpurwalla, N.D. 2007. "Betting on Residual Life: The Caveats of Conditioning." *Letters in Probability and Statistics* 77(12):1354-1361.

Singpurwalla, N.D., and J. Booker. 2004. "Membership Functions and Probability Measures of Fuzzy Sets." *Journal of the American Statistical Association* 99:867-877.

Singpurwalla, N.D., and A. Wilson. 2007. "Probability, Chance, and the Probability of Chance." *IIE Transactions.* Forthcoming.

von Neumann, J., and O. Morgenstern. 1944. *Theory of Games and Economic Behavior.* Princeton, N.J.: Princeton University Press.

Walley, P. 1991. *Statistical Reasoning with Imprecise Probabilities.* London: Chapman and Hall.

Watson, H.A. 1961. *Launch Control Safety Study.* Section VII, Vol. 1. Murray Hill, N.J.: Bell Laboratories.

Zadeh, L. 1979. "Possibility Theory and Soft Data Analysis, Memo." Technical Report UCB/ERL M79/66. Berkeley, Calif.: University of California.

Appendix H

Game Theory and Interdependencies

Geoffrey Heal, Ph.D.
Paul Garrett Professor of Public Policy and Business Responsibility
Columbia University, New York, New York

Howard Kunreuther, Ph.D.
Cecilia Yen Koo Professor of Decision Sciences and Public Policy
University of Pennsylvania, Philadelphia, Pennsylvania

There are certain bad events that can only occur once. Death is the obvious example: an individual's death is irreversible and unrepeatable. More mundane examples are bankruptcy, being struck off a professional register, and other discrete events. In addition there are other events that can in principle occur twice but that are so unlikely and/or so dreadful that one occurrence is all that can reasonably be considered. The events of September 11, 2001, are perhaps of this type. A set of coordinated anthrax attacks in several highly populated regions is another. The fact that such events are typically probabilistic, taken together with the fact that the risk that one agent faces is often determined in part by the behavior of others, gives a unique and hitherto unnoticed structure to the incentives that agents face to reduce their exposures to these risks.

The key point is that the incentive that any agent has to invest in risk-reduction measures depends on how he or she expects the others to behave in this respect. For cases where there are complementarities or positive externalities, if the agent thinks that they will not invest in security, then this reduces the incentive for the agent to do so. On the other hand, should the agent believe that they will invest in security, then it may be best for it to do so also. So there may be an equilibrium where no one invests in protection, even though all would be better off if they had incurred this cost. Yet this situation does not have the structure of a prisoner's dilemma game, even though it has some similarities.

A fundamental question that needs to be posed is "Do individuals and organizations invest in security to a degree that is adequate from either a private or social perspective?" In general the answer is no, for reasons that are described below.

NOTE: This appendix is based on material appearing in Heal and Kunreuther (2006).

COMMON FEATURES OF THE PROBLEM

There are several different versions of this problem of interdependencies, and all have certain features in common. In what follows a payoff is assumed to be discrete and binary. A bad event either occurs or does not, and that is the full range of possibilities. You die or you live. A firm is bankrupt or not. An anthrax attack is successful or not in a densely urban city. A plane crashes or not. Another feature common to these interdependent problems is that the risk faced by one agent depends on the actions taken by others—there are externalities. The risk of an airline's plane being blown up by a bomb depends on the thoroughness with which other airlines inspect bags that they transfer to this plane. The risk that an anthrax attack in an urban city is successful depends on the nature of our system for preventing, detecting, and responding to the threat of biological weapons.

Finally there is a stochastic element in all of these situations. In contrast to the standard prisoner's dilemma paradigm where the outcomes are specified with certainty, the interdependent security problem involves chance events. The question addressed is whether to invest in security when there is some probability, often a very small one, that there will be a catastrophic event that could be prevented or mitigated. The risk depends in part on the behavior of others in the system. The unfavorable outcome is discrete in that it either happens or does not.

IMPORTANCE OF PROBLEM STRUCTURE

These three factors—non-additivity of damages, dependence of risks on the actions of others, and uncertainty—are, as we shall see, sufficient to ensure that there can be equilibria at which there is underinvestment in risk-prevention measures. The precise degree of underinvestment depends on the nature of the problem. To illustrate the nature of interdependencies we focus on two examples: airline security and computer security. If an airline accepts baggage that contains a bomb, this need not damage one of its own planes: it may be transferred to another airline before it explodes. So in this framework one agent may transfer a risk fully to another. It may of course also receive a risk from another. There is a game of "pass the parcel" here. The music stops when the bomb explodes. It can only explode once so only one plane will be destroyed.

The structure of this game is quite different in the case of computer networks. Here it is commonly the case that if a virus (or hacker) enters the network through one weak point, it (or he or she) then has relatively easy access to the rest of the network and can damage all other computers as well as the entry machine (Kearns, 2005). In this case the bad outcome has a characteristic similar to a public good: its consumption is non-rivalrous. Its capacity to damage is not exhausted after it has inflicted damage once. A bomb, in contrast, has a limited capacity to inflict damage, and this capacity is exhausted after one incident.

The computer network problem is similar to what might happen in a bioterrorist attack such as anthrax or smallpox where it is possible for contamination to spread across individuals. Even if an individual or firm has taken protective actions, there is still some chance that it can be contaminated or infected by others who have not undertaken similar measures and hence are at risk. For example, if a person has been vaccinated or taken preventive medicine against a disease, he or she may still contract the illness from others who have the disease if the vaccine or medicine is not 100% effective. In these cases where there are complementarities or positive externalities created by an individual taking protective measures, there is more incentive for one unit to invest in protective measures if the other units have taken similar actions. In fact, investing in security is most effective if all elements of the system obtain protection and weak links may lead to suboptimal behavior by everyone.

In both cases, the airline and computer security problems, the incentives depend on what others do. Suppose that there are a large number of agents in the system. In Kunreuther and Heal (2003) we show that in the computer security problem, if none of the other machines are protected against viruses or hackers then the incentive for any agent to invest in protection approaches zero. For airline security, if no other airline has invested in baggage checking systems and there is a high probability that bags will be transferred from one airline to another, the expected benefits to any airline from this investment approaches 63% of what it would have been in the absence of contagion from others.

As we show below there can be a stable equilibrium where all agents choose not to invest in risk reduction measures, even though all would be better off if they did invest. An interesting property of some of these equilibria is the possibility of *tipping* as described by Schelling (1978). How can we ensure that if enough agents will invest in security that all the others will follow suit? In some cases there may be one agent occupying such a strategic position that if it changes from not investing to investing in protection, then all others will find it in their interests to do the same. And even if there is no single agent that can exert such leverage, there may be a small group. Obviously this finding has significant implications for policy-making. It suggests that there are some key players whom it is particularly important to persuade to manage risks carefully. Working with them may be a substitute for working with the population as a whole.

CHARACTERIZING THE PROBLEM: TWO-AGENT PROBLEM

We now set out formally the framework to study interdependent security (henceforth denoted IDS). Consider two identical airlines, A_1 and A_2, each having to choose whether or not to invest in a baggage screening system. Each faces a risk of a bomb exploding on its plane, causing a loss of L. There are two possible ways in which damage can occur: a bomb can explode either in a bag initially checked onto the airline's own plane or in a bag transferred from the other airline. The probability of a bomb exploding in luggage initially checked on a plane of an airline that has not invested in security is p. The expected loss from this event is pL. If the airline has invested in security precautions then this risk is assumed to be zero.

Even if an airline has invested in a baggage screening system there is still an additional risk of loss due to *contagion* from the other airline if it has not invested in security. The probability of a dangerous bag being accepted by one airline and then being transferred to the other is denoted by q. With respect to the chances of contagion, q is the likelihood that on any trip a dangerous bag is loaded onto the plane of one airline and is then transferred to another airline where it explodes. We assume that there is not enough time for an airline to examine the bags from another airline's plane before they are loaded onto its own plane.

These probabilities are interpreted as follows. On any given trip there is a probability p that an airline without a security system loads a bomb that explodes on one of its own planes. For the airline scenario, thorough scanning of baggage that an airline checks on its own plane will prevent damage from these bags, but there could still be an explosive in a bag transferred from another airline. There is thus an additional risk of loss due to contagion from another agent who has not invested in loss prevention, denoted by q. If there are $n \geq 2$ airlines, the probability per trip that this bag will be transferred from airline i to airline j is $q/(n-1)$. Note that the probability per trip that a bag placed on an airline without a security system will explode in the air is $p + q$.

We assume throughout that the damages that result from multiple security failures are no more severe than those resulting from a single failure. In other words, damages are not additive. In the airline baggage scenario, this amounts to an assumption that one act of terrorism is as serious as several. In reality, having two bombs explode on a plane is no more damaging than just one. The key issue is whether or not there is a failure, not how many failures there are. Indeed as the probabilities are so low, single occurrences are all that one can reasonably consider. One could think of the definition of a catastrophe as being an event so serious that it is difficult to imagine an alternative event with greater consequences. We focus first on the case of two airlines, each of which is denoted as an agent. This example presents the basic intuitions in a simple framework. We then turn to the multi-agent case.

To illustrate the framework in the context of a real-world event, consider the destruction of Pan Am flight 103 in 1988. In Malta terrorists checked a bag containing a bomb on Malta Airlines, which had minimal security procedures. The bag was transferred at Frankfurt to a Pan Am feeder line and then loaded onto Pan Am 103 in London's Heathrow Airport. The transferred piece of luggage was not inspected at either Frankfurt or London, the assumption in each airport being that it was inspected at the point of origin. The bomb was designed to explode above 28,000 feet, a height normally first attained on this route over the Atlantic Ocean. Failures in a peripheral part of the airline network, Malta, compromised the security of a flight leaving from a core hub, London.

Assume that each airline has two choices: to invest in baggage screening, *S*, or not to do so, *N*. Table H.1 shows the payoffs to the agents for the four possible outcomes.

Here *Y* is the income of each airline before any expenditure on security or any losses from the risks faced. The cost of investing in security is *c*. The rationale for these payoffs is straightforward. If both airlines invest in security, then each incurs a cost of *c* and faces no losses from damage so that their net incomes are $Y - c$. If A_1 invests and A_2 does not (top right entry) then A_1 incurs an investment cost of *c* and also runs the risk of a loss from damage emanating from A_2. The probability of A_2 contaminating A_1 is *q*, so that A_1's expected loss from damage originating elsewhere is qL. This cost represents the negative externality imposed by A_2 on A_1. In this case A_2 incurs no investment costs and faces no risk of contagion but does face the risk of damage originating at home, pL. The lower left payoffs are just the mirror image of these.

If neither airline invests, then both have an expected payoff of $Y - pL - (1 - p)qL$. The term pL here reflects the risk of damage originating at one's own airline. The term qL, showing the expected loss from damage originating at the other airline, is multiplied by $(1 - p)$ to reflect the assumption that the damage can only occur once. So the risk of contagion only matters to an airline when that airline does not suffer damage originating at home.

The conditions for investing in security to be a dominant strategy are that $c < pL$ and $c < p(1 - q)L$. The first constraint is exactly what one would expect if there were only a single airline: the cost of investing in security must be less than the expected loss. Adding a second airline tightens the constraint by reflecting the possibility of contagion. This possibility reduces the incentive to invest in security. Why? Because in isolation investment in security buys the airline complete freedom from risk. With the possibility of contagion it does not. Even after investment there remains a risk of damage emanating from the other airline. Investing in security buys you less when there is the possibility of contagion from others.

This solution concept is illustrated below with a numerical example. Suppose that $p = .2$, $q = .1$, $L = 1000$ and $c = 185$. The matrix in Table H.1 is now represented as Table H.2.

One can see that if A_2 has protection (*S*), then it is worthwhile for A_1 to also invest in security since its expected losses will be reduced by $pL = 200$ and it will only have to spend 185 on the security measure. However, if A_2 does not invest in security (*N*), then there is still a chance that A_1 will incur a loss. Hence the benefits of security to A_1 will only be $pL(1 - q) = 180$ which is less than the cost of the protective measure. So A_1 will *not* want to invest in protection. In other words, either both airlines invest in security or neither of them does so. These are the two Nash equilibria for this problem.

THE MULTI-AGENT IDS CASE

The results for the two-agent case carry over to the most general settings with some increase in complexity. In this section we review briefly the main features of the general cases, without providing detailed proofs of the results. These can be found in Kunreuther and Heal (2003).

There are two key points that emerge from the discussion of the general case with respect to the IDS problem. One is

TABLE H.1 Expected Costs Associated with Investing and Not Investing in Airline Security

		Airline 2 (A_2)	
		S	*N*
Airline 1 (A_1)	*S*	$Y - c, Y - c$	$Y - c - qL, Y - pL$
	N	$Y - pL, Y - c - qL$	$Y - [pL + (1 - p)qL]$, $Y - [pL + (1 - p)qL]$

NOTE: *S*, screening of baggage; *N*, no screening.

TABLE H.2 Expected Costs Associated with Investing and Not Investing in Airline Security: Illustrative Example

		Airline 2 (A_2)	
		S	*N*
Airline 1 (A_1)	*S*	$Y - 185, Y - 185$	$Y - 285, Y - 200$
	N	$Y - 200, Y - 285$	$Y - 280, Y - 280$

NOTE: *S*, screening of baggage; *N*, no screening.

that the main feature of the two-agent case carries over to n agents: the incentive that any agent faces to invest in security depends on how many other agents there are and on whether or not they are investing. Other agents who do not invest reduce the expected benefits from one's own protective actions and hence reduce an agent's incentive to invest.

Secondly there is a new possibility that emerges from the multi-agent case. There is the possibility of a *tipping* phenomenon.[1] In some cases there may be one firm occupying such a strategic position that if it changes from not investing to investing in protection, then all others will find it in their interests to follow suit. And even if there is no single firm that can exert such leverage, there may be a small group. Heal and Kunreuther (2007) show when this can happen and how to characterize the agents with great leverage. Obviously this point has considerable implications for policy-making. It suggests that there are some key players whom one needs to persuade to manage risks carefully.

EXTENDING THE ANALYSIS

The choice of whether to protect against events where there is interdependence between your actions and those of others raises a number of interesting theoretical and empirical questions. We mention some of these in this section.

Differential Costs and Risks

The nature of Nash equilibria for the problems considered above and the types of policy recommendations may change as one introduces differential costs across the agents who are considering whether or not to invest in security. Consider each airline deciding whether to invest in a baggage security *system*. In Heal and Kunreuther (2007) we have shown that if there are differential costs and/or risks between companies, we would expect to find some airlines investing in baggage security systems and others who would not. Furthermore, as we discussed above, the airline which creates the largest negative externalities for others should be encouraged to invest in protective behavior not only to reduce these losses but also to make it profitable for other airlines to follow suit, thus inducing tipping behavior.

Multi-Period and Dynamic Models

Deciding whether to invest in security normally involves multi-period considerations since there is an upfront investment cost that needs to be compared with the benefits over the life of the protective measure. An airline that invests in a baggage security system knows that this measure promises to offer benefits for a number of years. Hence one needs to discount these positive returns by an appropriate interest rate and specify the relevant time interval in determining whether or not to invest in these actions. There may be some uncertainty with respect to both of these parameters. From the point of view of dynamics, the decision to invest depends on how many others have taken similar actions. How do you get the process of investing in security started? Should one subsidize or provide extra benefits to those willing to be innovators in this regard to encourage others to take similar actions?

Endogenous Probabilities

The above analysis assumed that the risks faced by the airlines are independent of their own behavior. In reality if some airlines are known to be more security-conscious than others, they are presumably less likely to be terrorist targets. In this sense the problem of investing in security has similarities to the problem of theft protection: if a house announces that it has installed an alarm, then burglars are likely to turn to other houses as targets. In the case of airline security, terrorists are more likely to focus on targets that are less well protected. This is the phenomenon of displacement or substitution, documented in Sandler (2005). Keohane and Zeckhauser (2003) and Bier (2007) also consider the case of endogenous terrorist risks.

For the case of endogenous probabilities in the airline security problem, Heal and Kunreuther (2007) show that an airline is more likely to invest in security when probabilities are endogenous than when these probabilities are exogenous because of the increased likelihood of being a target when others invest in protection. In addition, if one makes the reasonable assumption that the total externality imposed on any non-investing firm decreases as the number of investing firms increases, then this should lead more firms to invest in protection. For both these reasons it should also now be easier for a coalition to tip the other firms into investing in security than if the probabilities were exogenous. Future research should examine how changes in endogenous probabilities impact on IDS solutions and the appropriate strategies for improving individual and social welfare.

Behavioral Considerations

The models discussed above all assumed that individuals made their decisions by comparing their expected benefits with and without protection to the costs of investing in security. This is a rational model of behavior. As pointed out in Chapter 2 of this report, there is a growing literature in behavioral economics that suggests that individuals make choices in ways that differ from the rational model of choice. With respect to protective measures there is evidence from controlled field studies and laboratory experiments that many individuals are not willing to invest in security for a number of reasons that include myopia, high discount rates and budget constraints (Kunreuther et al., 1998). In the models

[1]See Schelling (1978) for a characterization of a number of tipping problems.

considered above there were also no internal positive effects associated with protective measures. Many individuals invest in security to relieve anxiety and worry about what they perceive might happen to them or to others so as to gain peace of mind (Baron et al., 2000). A more realistic model of interdependent security that incorporated these behavioral factors as well as people's misperceptions of the risk may suggest a different set of policy recommendations than a rational model of choice.

FUTURE RESEARCH ON RISK MANAGEMENT STRATEGIES FOR IDS PROBLEMS

We conclude by suggesting a set of problems that involve interdependent security and suggesting the types of risk management strategies that could be explored for addressing them.

Types of Problems

The common features of IDS problems are the possibility that other agents can contaminate you and your inability to reduce this type of contagion through investing in security. You are thus discouraged from adopting protective measures when you know others have decided not to take this step. Here are some problems that fit into this category, some of which have been discussed in this paper:

- Investing in airline security
- Protecting against bioterrorist attacks
- Protecting against chemical and nuclear reactor accidents
- Making buildings more secure against attacks
- Investing in sprinkler systems to reduce the chance of a fire in one's apartment
- Making computer systems more secure against terrorist attacks
- Investing in protective measures for each part of an interconnected infrastructure system such as electricity, water or gas so that services can be provided to victims of a disaster

In each of these examples there are incentives for individual units or agents not to take protective measures but there are large potential losses to the unit making a decision (e.g., individual, organization, city) as well as to society. In the case of bioterrorism, if each unit takes protective action it will create positive externalities to others in the system and to society. Furthermore, the losses from these events are sufficiently high that they are considered to be non-additive. One can only get a specific disease once (e.g., smallpox, anthrax), an airplane can only be destroyed once; a building can only collapse once. You can only die once!

These IDS problems can be contrasted with others that do *not* have these features. One that is discussed in more detail in Kunreuther and Heal (2003) is theft protection where there are negative externalities to others from your taking protection. In the case of theft protection, if you install an alarm system that you announce publicly with a sign, the burglar will look for greener pastures to invade.[2]

Risk Management Strategies

For each IDS problem there are a range of risk management strategies that can be pursued by the private and public sectors for encouraging agents to invest in cost-effective protective measures.

- Collecting information on the risk and costs (e.g., constructing a scenario so that one can estimate p, q, L, and c with greater accuracy);
- Developing more accurate catastrophe models for examining the risk of terrorist attacks and other large-scale disasters;[3]
- Designing incentive systems (e.g., subsidies or taxes) to encourage investment by agents in protective measures;
- Developing insurance programs for encouraging investment in protective measures when firms are faced with contagion;
- Structuring the liability system to deal with the contagion effects of IDS;
- Carefully designed standards (e.g., building codes for high-rises to withstand future terrorist attacks) that are well enforced through mechanisms such as third-party inspections;
- Introducing federal reinsurance or state-operated pools to provide protection against future losses from terrorist attacks to supplement private terrorist insurance.

It may be desirable to integrate several of these measures through public-private risk management partnerships. For example, banks and financial institutions could require that firms adopt security measures as a condition for a loan or mortgage. To ensure that these measures are adopted there may be a need for third party inspections or audits by the private sector. Firms who reduce their risks can be rewarded through lower insurance premiums. If there are federal or state reinsurance pools at reasonable prices to cover large losses from a future terrorist attack, then private insurers may be able to provide terrorist coverage at affordable premiums.

---

[2]One could make a similar argument with respect to cities taking protective measures against bioterrorism. For example, if certain cities were equipped with sensors to detect biological attacks, the terrorist might focus his or her attention on those urban areas that did not have this form of protection.

[3]For more details on the challenges in developing catastrophe models and appropriate strategies for dealing with them, see Grossi and Kunreuther (2005).

REFERENCES

Baron, J., J. Hershey, and H. Kunreuther. 2000. "Determinants of Priority for Risk Reduction: The Role of Worry." *Risk Analysis* 20(4):413-427.

Bier, V. 2007. "Choosing What to Protect." *Risk Analysis* 27 (June):607-620.

Grossi, P., and H. Kunreuther. 2005. *Catastrophe Modeling: A New Approach to Managing Risk.* New York: Springer.

Heal, G., and H. Kunreuther. 2006. "You Can Only Die Once: Interdependent Security in an Uncertain World." In *The Economic Impacts of Terrorist Attacks*, H.W. Richardson, P. Gordon, and J.E. Moore, III (eds.). Northampton, Mass.: Edward Elgar.

Heal, G., and H. Kunreuther. 2007. "Modeling Interdependent Risks." *Risk Analysis* 27(3):621-633.

Kearns, M. 2005. "Economics, Computer Science and Policy." *Issues in Science and Technology*, Winter: pp. 37-47.

Keohane, N., and R. Zeckhauser. 2003. "The Ecology of Terror Defense." *Journal of Risk and Uncertainty. Special Issue on Terrorist Risks* 26(2/3):201-229.

Kunreuther, H., and G. Heal. 2003. "Interdependent Security." *Journal of Risk and Uncertainty, Special Issue on Terrorist Risks* 26(2/3):231-249.

Kunreuther, H., A. Onculer, and P. Slovic. 1998. "Time Insensitivity for Protective Measures." *Journal of Risk and Uncertainty* 16(3):279-299.

Sandler, T. 2005. "Collective Action and Transnational Terrorism." *The World Economy* 26(6):779-802.

Schelling, T. 1978. *Micromotives and Macrobehavior.* New York: Norton.

Appendix I

Review of BTRA Modeling

Alan R. Washburn, Ph.D.
Distinguished Professor Emeritus of Operations Research
Naval Postgraduate School, Monterey, California

July 10, 2007

MEMORANDUM FOR THE NATIONAL ACADEMY OF SCIENCES (NAS)

Review of the Department of Homeland Security (2006) work on bioterrorism.

Background. The Department of Homeland Security (DHS) has produced a 2006 bioterrorism study, and is working on subsequent versions. DHS has asked NAS to assess the 2006 work, which I will refer to hereafter as "the 2006 work." I have become acquainted with the work through contacts with the NAS committee, and have been invited to provide a review. This is the review. It is intended for a scientific audience, so I will not hesitate to use the language of probability in describing what I think was done in 2006, or in how things might be handled differently in the future. Random variables are uppercase symbols, P() and E() are the probability and expected value functions, respectively.

My Qualifications. After working five years for the Boeing Company, I joined the Operations Research faculty at the Naval Postgraduate School in 1970, where I did the usual academic things until retiring in 2006. My teaching includes probability and decision theory, which are relevant here. See my resume at http://www.nps.navy.mil/orfacpag/resumePages/washbu.htm for details. I have no biological or medical qualifications. My acquaintance with the work is mainly through the references listed at the end of this review.

Event Trees. The fundamental idea behind the 2006 work is an event tree. As I will use the term in this review, an event tree is a branching structure whose root corresponds to the assertion that some event has occurred, the event in this case being what I will call an "incident." The tree branches repeatedly until a "scenario" is encountered, at which point one will find a probability distribution that determines the consequence of the incident, a random variable that I will call Y. I think of consequences as being "lives lost," but any other scalar measure would do. Each node of the tree has a set of successor arcs, and there is a given probability distribution over these arcs. One can imagine starting at the root and randomly selecting an arc at each node encountered until finally the consequence is determined. In addition to Y, the event tree involved in the 2006 work is such that every path from root to consequence also defines two other random variables:

- A, the biological agent, one of 28 possibilities, and
- S, the scenario.

The scenario might be null in the sense that Y is 0 because the incident is terminated prematurely, but is nonetheless always defined.

DHS determines the consequence distributions through Monte Carlo simulation based on expert input. The results are collected into decade-width histograms. I will not comment further on the methodology for producing the consequence distributions, since I have not examined it in detail.

DHS has modified the above definition of an event tree in three senses. One is that the initial branches from the root are rates, rather than probabilities. Call the rate on branch i λ_i, and let the sum of all of these rates be λ. If one interprets these rates as independent Poisson rates of the various kinds of incident, then it is equivalent to think of incidents as occurring in a Poisson process with rate λ, with each incident being of type i with probability λ_i/λ. These ratios can be the first set of branch probabilities, so this is all equivalent to the standard event tree definition, except that we must remember that incidents occur at the given rate λ. This first modification is thus of little import.

The second modification is that an incident might involve multiple attacks, each with separate consequences. This is a more significant modification, and will be discussed separately below.

The third and most significant modification is that the branching probabilities (DHS on occasion also calls them "branch fractions") are not fixed, but are instead themselves determined by sampling from beta distributions provided indirectly by Subject Matter Experts (SMEs). Let θ be the collection of branching probabilities. In each incident we therefore observe (θ, *A*, *S*, *Y*), with θ determining the event tree for the other three random variables. This modification will also be discussed separately below.

The Second Modification: Repeated Attacks per Incident. The vision is that a cell or group of terrorists will not plan a single attack, but will plan to continue to attack until interrupted, with the entire group of attacks constituting an incident. The effect of this is to change the distribution of consequences of an incident, since a successful attack will be accompanied by afterattacks, the number of which I will call *X*. I believe that the formula used for calculating E(*X*) is incorrect. Specifically, let λ' be the probability that any one of the afterattacks will succeed, assume that afterattacks continue until one of them fails, and assume that the failed afterattack terminates the process and itself has no consequences. Then the average value of *X* is $E(X) = \lambda'/(1 - \lambda')$, the mean of a geometric-type random variable. This is not the formula in use. Using the correct formula would be a simple enough change, but I believe the numerical effect might be significant.

Other changes may also be necessary to implement the original vision. If the afterattacks all have independent consequences, then the distribution of total consequences is the (1 + *X*)-fold convolution of the consequence distribution, a complicated operation that I see no evidence of. The documentation is mute on what is actually assumed about the independence of after attacks, and on how the E(*X*) computation is actually used. Simply scaling up the consequences of one attack by the factor (1 + E(*X*)) is correct on the average, regardless of independence assumptions, but will not give the correct distribution of total consequences.

The Third Modification: "Random Probabilities." DHS has accommodated SME uncertainty by allowing the branch probabilities themselves to be random quantities, with the SMEs merely agreeing to a distribution for each probability, rather than a specific number. I will refer to each of these probability distributions as a "marginal" for its branch. If a node has *N* branches, the experts contribute *N* marginals, one for each branch. Except at the root, these marginals are all beta distributions on the interval [0 1], and each therefore has two parameters, alpha (α) and beta (β). Each of these distributions has a mean, and since the probabilities themselves must sum over the branches to 1, the same thing must logically be true of the means. The same need not be true of the SME inputs, but DHS seems to have disciplined the elicitation process so that the SME marginal means actually do sum to 1. That is true in all of the data that I have seen.

However, summing to 1 is not sufficient for the SME marginals to be meaningful. This is most obvious when $N = 2$. If the first branch has probability *A*, then the second must have probability $1 - A$, and therefore the second probability distribution has no choice but to be the mirror image of the first. If the experts feel that the first marginal has $\alpha = 1$ and $\beta = 1$, while the second has $\alpha = 2$ and $\beta = 2$, then we must explain to the experts that what they are saying is meaningless, even though both marginals have a mean of 0.5. The second marginal has no choice but to be the mirror image of the first, and must therefore *be* the first, by symmetry. Any other possibility is literally meaningless, since there is no pair of random variables (A_1, A_2) such that A_i has the *i*th marginal distribution and also $A_1 + A_2$ is always exactly 1.

I think DHS recognizes the difficulty when $N = 2$, and has basically fixed it in that case by asking the SMEs for only one marginal, but the same difficulty is present for $N > 2$, and has not been fixed. The sampling procedure offered on page C-81 of Department of Homeland Security (2006) will reliably produce probabilities $A_1, \ldots, A_N$ that sum to 1, and which are correct on the average, but they do not have the marginal beta distributions given by the SMEs. This is most obvious in the case of the last branch, since the *N*th marginal is never used in the sampling process, but I believe that the marginal distribution is correct only for the first branch.

There is a multivariable distribution (the Dirichlet distribution) whose marginals are all beta distributions, but the Dirichlet distribution has only $N + 1$ parameters. The SME marginals require 2*N*, in total, so the Dirichlet distribution is not a satisfactory joint distribution for $A_1, \ldots, A_N$.

Estimation of the Spread in Agent-Damage Charts. I have defined *Y* to be the consequence and *A* to be the agent. Define Y_a to be the consequence if $A = a$, or otherwise 0, so that the 28 random variables Y_a sum to *Y*. Most of the DHS output deals with the random variable $E(Y_a | \theta)$, the expected consequence contribution from agent *a*, given the sampled branch probabilities θ. This quantity is random only because of its dependence on θ, the natural variability of Y_a having been averaged out. A sample $E(Y_a | \theta_j)$, j = 1,…, 500 is produced by Latin Hypercube Sampling (LHS) of the branch probabilities, each sample including the standard average risk computations for the event tree. A sample mean estimate $\hat{Y}_a$ of $E(Y_a)$ is then made by $\hat{Y}_a = (1/500)\sum_{j=1}^{500} E(Y_a | \theta_j)$. The agents are then sorted in order of decreasing sample mean, and displayed in what I will call "agent-damage" charts showing the expected values and spreads as a function of agent. The sample means are normalized before being displayed, probably by forcing them to sum to 1. The normalization destroys information that is relevant to the decisions being made. I do not know the motivation for doing so.

The spreads display the epistemic variability due to SME uncertainty about θ, but suppress all of the aleatoric variability implied by the event tree. If there were no uncertainty about θ, all of the spreads would collapse to a single point (the mean) for each agent. I am not sure how the variability displayed in agent-damage charts is supposed to relate to decision making, but I guess that the graphs are intended to support conclusions such as the following: "I know that the mean damage for agent 1 is larger then the mean damage for agent 2, but I still think that we ought to spend our money defending against agent 2 because of its high associated variability. Even a small prospect of the high damages associated with agent 2 is not acceptable." If that is the kind of logic that the agent-damage charts are intended to support, then they should include aleatoric variability. Without it, the spreads associated with each agent are too small. This issue affects infectious agents more than the other kind, since infectious diseases will have especially high damage variances.

The agent-damage charts are intended for a high level of decision-making audience, and devote considerable space (one of the two available dimensions) to showing the spread associated with each agent. Without the need to show spread, they could be replaced by bar charts or simple tables. If spread is important enough to be displayed, then it ought to be displayed in a manner that facilitates good decisions. I doubt that that is currently the case.

Even without the aleatoric issue, I still have concerns about the spread that is displayed. The object ought to be to display the mean and fractiles (the spread) of the random variable $E(Y_a \mid \theta)$ for each value of a. The mean of $E(Y_a \mid \theta)$ is simply $E(Y_a)$ by the conditional expectation theorem, and is estimated by $\hat{Y}_a$. DHS claims graphically that the LHS sample fractiles are also the fractiles of the random variable $E(Y_a \mid \theta)$. I suspect that this claim is false. LHS is basically a variance reduction technique that makes the variance of $\hat{Y}_a$ smaller than it would be with ordinary sampling. While this effect is welcome, LHS also has an unpredictable effect on variability. The spread that is shown for each agent may not be a good estimate of the spread of the random variable $E(Y_a \mid \theta)$.

One final point on estimation. As long as there is no dependence between the branch probabilities at different nodes, as there is not in the 2006 work, it is characteristic of an event tree that $P(Y_a \leq y) = E(P(Y_a \leq y \mid \theta)) = P(Y_a \leq y \mid E(\theta))$. The first equality is due to the conditional expectation theorem, and the second is because no event tree probability enters more than once into calculating the probability of any scenario. In other words, all information pertinent to the distribution of Y_a could be obtained without sampling error by simply replacing the marginal branch distributions by their means. This information includes $E(Y_a)$, which is currently being estimated (with sampling error) by $\hat{Y}_a$. (Note added in June 2007. Let me expand the notation to clarify this final point, since it has caused some confusion. Let $\theta = (Q_1, \ldots, Q_n)$, where n is the number of nodes and Q_i is the collection of branch probabilities at node i. Also let Q_{ij} be the jth branch probability at node i. In the sampling procedure used by DHS to obtain θ, Q_{ij} and Q_{kl} are independent random variables as long as i and j are not the same, which is all that is required for my conclusion to be true. While it is certainly true that the branches chosen at nodes i and j are in general dependent, the branch *probabilities* are not.)

Use of SMEs. It is inevitable in a project like this that probabilities will have to be obtained from Subject Matter Experts, rather than experimentation. The important thing is that the SMEs at least know what they are estimating, and that estimates be used correctly once they are obtained. I have already mentioned that SME estimates of the marginal branch distributions are not reproduced by the sampling procedure. Another concern is at the third stage of the event tree, where SMEs are asked to deal with agent selection. At that stage there are $4 \times 8 = 32$ nodes in the event tree where an agent might be selected, each of which has 28 branches. I can certainly understand DHS's reluctance to conduct 896 interviews with SMEs, each to determine one of the needed beta distributions. Some kind of a shortcut is needed, but I wonder whether the one adopted is a good one. The SMEs are first asked to determine an "input regarding known preferences of terrorists" for each agent. If I were an SME and somebody asked me to determine the quoted expression for agent a, I would announce my estimate of $P(A = a)$, the probability that agent a is actually selected in an incident. Given all of these SME inputs, DHS then goes over the 896 branches, some of which have a logical 0 for the agent, and assigns probabilities using the rule that the probability is either 0 or else proportional to the SME's agent input, the proportionality constant being selected in each of the 32 cases so that the probabilities sum to 1. My objections are that

- The quoted expression above does not make it clear that the SME input is supposed to be $P(A = a)$. There is a danger of every SME making a different interpretation of what is being asked for.
- If the SME does input the probabilities $P(A = a)$, and if DHS applies the shortcut procedure to fill out the third stage of the event tree, and if the probabilities of the 28 agents are then computed from the tree, they will not necessarily agree with the SME's inputs. This would be true even without my next objection.
- The SME's inputs are subsequently modified by various formulas involving agent lethality, etc. What is an SME who is already acquainted with agent lethality to think of this? Should he adjust his input so that the net result of all this computation is the number that he wanted in the first place? If one is going to elicit SME inputs on probabilities, then it seems to me that one ought to use them as they are intended.

Given that the agent probabilities strongly influence the agent-damage charts, the procedure for eliciting and using them should be an object of concern in future work.

Tree Flipping? The process described earlier for generating agent-damage charts may not be a correct statement of what DHS actually did in 2006. The DHS documentation in several places, after describing a single event tree with 17 ranks, states that a separate analysis was actually done for each agent (paragraph C.3.4.2 of Department of Homeland Security [2006], for example). Now, it is possible to end up with the single-tree analysis described earlier by doing that. The essential step is to first calculate $P(A = a)$ for each agent, and then make a new tree where the agent is selected at the root, with the agent selection probabilities on the 28 branches from the root. The second and third ranks of the tree would then be what were originally the first and second, with new probabilities as computed by Bayes' theorem, and the rest of the tree would be unchanged. Since the agent is at the root of the resulting "flipped" tree, using the flipped tree is in effect doing a separate analysis for each agent. The flipped tree would lead to the same earlier described agent-damage charts—the two trees are stochastically equivalent. But I don't see the motivation for doing all this extra work in flipping the tree, and I have some concerns about whether the flipping operation was actually done correctly, or done at all.

One concern is that the thing being manipulated is not an ordinary event tree, and there is no reason to expect that beta distributions will remain beta distributions in the flipping process. Of course, the flipping could occur after the tree is instantiated in each of the 500 replications, but that would get to be a lot of work. I doubt if that has been the case.

The documentation is mute about the tree flipping process. I can only hope that the method actually used for producing agent-damage charts is equivalent to analyzing the single event tree as described above.

Suggestions. My main suggestion for future work is that distributions for branch probabilities be abandoned in favor of direct branch probabilities, as in a standard event tree. In other words, keep it simple. SMEs will not be comfortable expressing definite values for the probabilities, but then they are probably not comfortable with expressing definite values for α and β, either. Most people are simply not comfortable quantifying uncertainty. There is very little to be gained by including epistemic uncertainty about the branch probabilities in an analysis like this, and much to be lost in terms of complication. Epistemic uncertainty is not even discussed in most decision theory textbooks. Standard software for handling decision trees would become applicable (event trees are just a special case where there are no decisions) if epistemic uncertainty were not present. There is also standard software for handling influence diagrams, which ought to be considered as an alternative to decision trees. Influence diagram software is sometimes used diagnostically, which might be of use in bioterrorism. One might observe that the agent is known to be anthrax, for example, and instantly recompute the target probabilities based on that known condition.

Another suggestion is to examine the potential for optimization. Given that the basic problem is how to spend money to reduce risk, it is too bad that a problem that simple in structure cannot be posed formally. It is possible that some actions that we might take would be effective for *all* contagious diseases. This should make them attractive, but the low rank of most contagious diseases individually in the agent-damage charts tends to suppress their attractiveness.

My last suggestion is to report future results in a scientific fashion that can be reviewed by scientists. English is a notoriously imprecise language for describing operations involving chance, so I have repeatedly struggled to understand what was actually done in making my way through the references. As a result, I may well have misinterpreted something above that I hope DHS will correct. If I were reviewing the 2006 work for a journal, my first act would be to send the material back to the authors with a request that it be written up using mathematics embedded in English, instead of just English. I know that DHS has to communicate complicated ideas about risk to laypeople. That task should be in addition to reporting the results scientifically, not a replacement for it.

In summary, my opinion is that the 2006 DHS methodology is not yet the "rigorous and technically sound methodology" demanded by the 2004 Homeland Security Presidential Directive 10: *Biodefense for the 21st Century*. Let me also add that I consider the report as a whole to be a remarkable accomplishment, given the magnitude of the task and the time available to do it.

References. Materials that I have examined before writing this review include the following:

Department of Homeland Security. 2006. *Bioterrorism Risk Assessment.* Biological Threat Characterization Center of the National Biodefense Analysis and Countermeasures Center. Fort Detrick, Md.

I have also examined various drafts of the following:

Department of Homeland Security. 2007. "A Lexicon of Risk Terminology and Methodological Description of the DHS Bioterrorism Risk Assessment." April 16.

Of all the documents, this last one comes closest to the technical appendix that I recommend. It has been of considerable use to me, but even it does not address tree flipping.

Appendix J

Reprinted Interim Report

Interim Report on Methodological Improvements to the Department of Homeland Security's Biological Agent Risk Analysis

Committee on Methodological Improvements to the
Department of Homeland Security's
Biological Agent Risk Analysis

Board on Mathematical Sciences and Their Applications

Division on Engineering and Physical Sciences

NATIONAL RESEARCH COUNCIL
OF THE NATIONAL ACADEMIES

THE NATIONAL ACADEMIES PRESS
Washington, D.C.
www.nap.edu

THE NATIONAL ACADEMIES PRESS 500 Fifth Street, N.W. Washington, DC 20001

NOTICE: The project that is the subject of this report was approved by the Governing Board of the National Research Council, whose members are drawn from the councils of the National Academy of Sciences, the National Academy of Engineering, and the Institute of Medicine. The members of the committee responsible for the report were chosen for their special competences and with regard for appropriate balance.

This study was supported by Contract No. HSHQDC-06-C-00046 between the National Academy of Sciences and the Department of Homeland Security. Any opinions, findings, conclusions, or recommendations expressed in this publication are those of the author(s) and do not necessarily reflect the views of the organizations or agencies that provided support for the project.

Additional copies of this report are available from the National Academies Press, 500 Fifth Street, N.W., Lockbox 285, Washington, DC 20055; (800) 624-6242 or (202) 334-3313 (in the Washington metropolitan area); Internet, http://www.nap.edu.

Printed in the United States of America

THE NATIONAL ACADEMIES

Advisers to the Nation on Science, Engineering, and Medicine

The **National Academy of Sciences** is a private, nonprofit, self-perpetuating society of distinguished scholars engaged in scientific and engineering research, dedicated to the furtherance of science and technology and to their use for the general welfare. Upon the authority of the charter granted to it by the Congress in 1863, the Academy has a mandate that requires it to advise the federal government on scientific and technical matters. Dr. Ralph J. Cicerone is president of the National Academy of Sciences.

The **National Academy of Engineering** was established in 1964, under the charter of the National Academy of Sciences, as a parallel organization of outstanding engineers. It is autonomous in its administration and in the selection of its members, sharing with the National Academy of Sciences the responsibility for advising the federal government. The National Academy of Engineering also sponsors engineering programs aimed at meeting national needs, encourages education and research, and recognizes the superior achievements of engineers. Dr. Wm. A.Wulf is president of the National Academy of Engineering.

The **Institute of Medicine** was established in 1970 by the National Academy of Sciences to secure the services of eminent members of appropriate professions in the examination of policy matters pertaining to the health of the public. The Institute acts under the responsibility given to the National Academy of Sciences by its congressional charter to be an adviser to the federal government and, upon its own initiative, to identify issues of medical care, research, and education. Dr. Harvey V. Fineberg is president of the Institute of Medicine.

The **National Research Council** was organized by the National Academy of Sciences in 1916 to associate the broad community of science and technology with the Academy's purposes of furthering knowledge and advising the federal government. Functioning in accordance with general policies determined by the Academy, the Council has become the principal operating agency of both the National Academy of Sciences and the National Academy of Engineering in providing services to the government, the public, and the scientific and engineering communities. The Council is administered jointly by both Academies and the Institute of Medicine. Dr. Ralph J. Cicerone and Dr. Wm. A. Wulf are chair and vice chair, respectively, of the National Research Council.

www.national-academies.org

COMMITTEE ON METHODOLOGICAL IMPROVEMENTS TO THE DEPARTMENT OF HOMELAND SECURITY'S BIOLOGICAL AGENT RISK ANALYSIS

GREGORY S. PARNELL, U.S. Military Academy, *Chair*
DAVID BANKS, Duke University
LUCIANA BORIO, University of Pittsburgh
GERALD BROWN, Naval Postgraduate School
L. ANTHONY COX, JR., Cox Associates
JOHN GANNON, BAE Systems
ERIC HARVILL, Pennsylvania State University
HOWARD KUNREUTHER, University of Pennsylvania
STEPHEN MORSE, Columbia University
MARGUERITE PAPPAIOANOU, University of Minnesota
STEPHEN POLLOCK, University of Michigan
NOZER SINGPURWALLA, George Washington University
ALYSON WILSON, Los Alamos National Laboratory

Staff

SCOTT WEIDMAN, Director, Board on Mathematical Sciences and Their Applications
NEAL GLASSMAN, Senior Staff Officer, Board on Mathematical Sciences and Their Applications
KERRY BRENNER, Senior Staff Officer, Board on Life Sciences
BARBARA WRIGHT, Administrative Assistant

For more information on BMSA, see its Web site at http://www7.nationalacademies.org/bms/, write to BMSA, National Research Council, 500 Fifth Street, N.W., Washington, DC 20001, call (202) 334-2421, or send e-mail to bms@nas.edu.

Acknowledgments

This report has been reviewed in draft form by individuals chosen for their diverse perspectives and technical expertise, in accordance with procedures approved by the National Research Council's Report Review Committee. The purpose of this independent review is to provide candid and critical comments that will assist the institution in making its published report as sound as possible and to ensure that the report meets institutional standards for objectivity, evidence, and responsiveness to the study charge. The review comments and draft manuscript remain confidential to protect the integrity of the deliberative process. We wish to thank the following individuals for their review of this report:

John Bailar III, University of Chicago,
Gerald Dinneen, Lexington, Massachusetts,
Randall Larsen, The Institute for Homeland Security,
Stephen Robinson, University of Wisconsin,
Harvey Rubin, University of Pennsylvania, and
Lawrence Wein, Stanford University.

Although the reviewers listed above have provided many constructive comments and suggestions, they were not asked to endorse the conclusions or recommendations nor did they see the final draft of the report before its release. The review of this report was overseen by Frank Stillinger, Princeton University. Appointed by the National Research Council, he was responsible for making certain that an independent examination of this report was carried out in accordance with institutional procedures and that all review comments were carefully considered. Responsibility for the final content of this report rests entirely with the authoring committee and the institution.

The committee also acknowledges the valuable contribution of the following individuals, who provided input at the meeting on which this interim report is based:

James Petro, White House Homeland Security Council,
Adam Rose, Pennsylvania State University,
Detlof von Winterfeldt, University of Southern California, and
Staff of the Battelle Memorial Institute, Columbus, Ohio.

Contents

Executive Summary

In recognition of potential bioterrorist threats, President George W. Bush issued Homeland Security Presidential Directive 10 (HSPD10), "Biodefense for the 21st Century," on April 28, 2004.[1] This directive, as well as the National Strategy for Homeland Security,[2] published by the White House Office of Homeland Security in 2002, required assessments of the biological weapons threat to the nation and assigned the Department of Homeland Security (DHS) responsibility for conducting these assessments, in coordination with other appropriate federal departments and agencies. The first DHS bioterrorism risk assessment was completed on January 31, 2006, and the report documenting the assessment was published on October 1, 2006.[3]

THE COMMITTEE'S PRELIMINARY ASSESSMENT

The National Research Council (NRC) was asked by DHS to carry out a study to recommend improvements to the methodology used for DHS's first bioterrorism risk assessment. The NRC study will issue two reports: interim (this report), focused on near-term improvements that can begin in federal Fiscal Year 2007 (FY2007), and final, to recommend longer-term improvements.

On August 28-29, 2006, the NRC Committee on Methodological Improvements to the Department of Homeland Security's Biological Agent Risk Analysis met with representatives of DHS, its National Biodefense Analysis and Countermeasures Center (NBACC), Battelle Memorial Institute, the White House Homeland Security Council, and the Homeland Security Center for Risk and Economic Analysis of Terrorism Events (CREATE). The briefings at this meeting described a probabilistic risk assessment (PRA) of 28 bioagents. For each of the 28 pathogens, it used a 17-step event-tree analysis of paths (sequences of events and actions) that could lead to the deliberate exposure of civilian populations. The recommendations and discussion below are based solely on those briefings; DHS's bioterrorism risk assessment was not made available to the committee in time for this interim report.

[1]Homeland Security Presidential Directive 10, "Biodefense for the 21st Century," April 28, 2004, available at http://www.fas.org/irp/offdocs/nspd/hspd-10.html. Accessed Nov. 1, 2006.

[2]See www.dhs.gov/xlibrary/assets/nat_strategy_hls.pdf. Accessed Nov. 1, 2006.

[3]*Bioterrorism Risk Assessment.* 2006. Biological Threat Characterization Center of the National Biodefense Analysis and Countermeasure Center. Washington, D.C.

This interim report provides DHS with overall near-term guidance and direction for the further development of its risk analysis models. The committee's final report will address longer-term issues in the development of risk analysis capabilities for DHS. Because the topics discussed here will be studied in more depth and with a view toward the longer term, the committee's final report will be more detailed and may modify the conclusions presented here. The committee is confident, however, that the recommendations included in this interim report are appropriate and necessary in the near term.

The committee recognizes that the development of this comprehensive suite of techniques used for the PRA is a logical extension of previous risk analysis methods used for natural and technological hazards and engineering design.[4] The implementation of the selected PRA framework appears, for the most part, to be consistent with well-accepted practice in other fields of risk analysis such as nuclear reactor safety and chemical safety. The committee also notes that DHS and its NBACC have sought ways to refine and improve this new capability.

THE COMMITTEE'S INTERIM RECOMMENDATIONS FOR FY2007

Based on its August 28-29, 2006, briefings, the committee's main concerns are about the overall purpose and directions of DHS's risk analysis, the challenges involved in structuring and predicting the actions of determined adversaries, and the need to provide policy makers with a sound foundation for DHS's ongoing risk analyses. Following are three critical interim recommendations.

Recommendation 1: DHS should establish a clear statement of the long-term purposes of its bioterrorism risk analysis.

A clear statement of the long-term purposes of the bioterrorism risk analysis is needed to enunciate how it can serve as a tool to inform risk assessment, risk perception, and especially risk-management decision making. Criteria and measures should be specified for assessing how well these purposes are achieved. Key issues to be addressed by such a statement should include the following: who the key stakeholders are; what their short- and long-term values, goals, and objectives are; how these values, goals, and objectives change over time; how the stakeholders perceive the risks; how they can communicate their concerns about these risks more effectively; and what they need from the risk assessment in order to make better (more effective, confident, rational, and defensible) resource-allocation decisions. Other important issues are who should perform the analyses (contractors, government, both) and how DHS should incorporate new information into the analyses so that its assessments are updated in a timely fashion.

Recommendation 2: DHS should improve its analysis of intelligent adversaries.

Event-tree methodology was not developed to model the possible actions of intelligent adversaries. Traditional event-probability assessment and elicitation techniques for these assessments are not sufficient for modeling the actions of intelligent adversaries made in response to their opponents' defensive actions and/or in response to initial successes or failures in their own plan execution. Alternative techniques—including red teams (i.e., individuals, including both technologists and those with experience in targeting and strat-

[4]See, for instance, http://www7.nationalacademies.org/aseb/stamatelatos_nasa_presentation.pdf and http://www.ans.org/pubs/magazines/nn/docs/2000-3-2.pdf. Accessed Nov. 1, 2006.

egy, whose purpose is to simulate adversarial decision making) and attack-preference, decision-tree, attack-tree, or attack-graph models[5]—might be more suitable to complement elicitation.

Recommendation 3: DHS should increase its risk analysis methodology's emphasis on risk management.

It is unclear how the event-tree probabilistic risk assessment will support DHS's design and evaluation of alternative risk management strategies. The computational engine being developed by Battelle does not permit, let alone encourage, risk managers to explore "if resource allocation, then probable consequence" scenarios for evaluating alternative risk management strategies.[6] DHS needs to determine how strategies involving specific investments of resources in protection and countermeasures translate to changes in risk and impact terrorist plans and actions. Moreover, the model should have an interface and visualization component that makes its results and limitations easier to understand and be used by decision makers.

The committee encourages DHS to continue to build on, refine, and improve the probabilistic risk assessment foundation already laid down. The committee will continue to pursue these and additional topics in its review over the coming year.

---

[5]Attack trees and attack graphs are modeling techniques for understanding risk in complex situations. Both are graphical representations showing all ways to attack or damage a system. Decision trees are event trees with decisions represented as possible events. Attack-preference models examine decisions from the viewpoint of the attacker rather than the defender. See http://csdl2.computer.org/persagen/DLAbsToc.jsp?resourcePath=/dl/proceedings/&toc=comp/proceedings/itcc/2004/2108/01/2108toc.xml&DOI=10.1109/ITCC.2004.1286496. Accessed Nov. 1, 2006.

[6]The DHS methodology, as reflected in software, actually does allow changes in assumptions; but this must be done through an analyst and would require a significant time delay and limit the range of alternatives that could be examined.

Methodological Improvements to the Department of Homeland Security's Biological Agent Risk Analysis

BACKGROUND

In recognition of potential bioterrorist threats, President George W. Bush issued Homeland Security Presidential Directive 10 (HSPD10), "Biodefense for the 21st Century,"[1] on April 28, 2004. The directive requires assessments of the biological weapons threat to the nation:

> Another critical element of our biodefense policy is the development of periodic assessments of the evolving biological weapons threat. First, the United States requires a continuous, formal process for conducting routine capabilities assessments to guide prioritization of our on-going investments in biodefense-related research, development, planning, and preparedness. These assessments will be tailored to meet the requirements in each of these areas. Second, the United States requires a periodic senior-level policy net assessment that evaluates progress in implementing this policy, identifies continuing gaps or vulnerabilities in our biodefense posture, and makes recommendations for re-balancing and refining investments among the pillars of our overall biodefense policy. The Department of Homeland Security, in coordination with other appropriate Federal departments and agencies, will be responsible for conducting these assessments.[2]

The first Department of Homeland Security bioterrorism risk assessment was completed on January 31, 2006, and the report documenting the analysis was published on October 1, 2006.[3] This assessment and report implemented the requirement of the National Strategy for Homeland Security,[4] issued in July 2002 by the Office of Homeland Security, and of HSPD10 for DHS to assess the biological weapons threat in coordination with other appropriate federal departments and agencies. At DHS's request, the National Research Council (NRC) established the Committee on Methodological Improvements to the Department of Homeland Security's Biological Agent Risk Analysis to provide a review, via two reports (interim and final), of the methodology used in DHS's report.

The committee's first meeting was held at the National Academies' Keck Center in Washington, D.C., on August 28-29, 2006. The appendix contains the agenda for that meet-

[1]Available at http://www.fas.org/irp/offdocs/nspd/hspd-10.html. Accessed Nov. 1, 2006.

[2]Available at http://www.fas.org/irp/offdocs/nspd/hspd-10.html. Accessed Nov. 1, 2006.

[3]*Bioterrorism Risk Assessment.* 2006. Biological Threat Characterization Center of the National Biodefense Analysis and Countermeasure Center. Washington, D.C.

[4]See www.dhs.gov/xlibrary/assets/nat_strategy_hls.pdf. Accessed Nov. 1, 2006.

ing. The committee heard and discussed presentations regarding risk analysis for biological pathogens by representatives of DHS, its National Biodefense Analysis and Countermeasures Center (NBACC), Battelle Memorial Institute, the White House Homeland Security Council, and the Homeland Security Center for Risk and Economic Analysis of Terrorism Events (CREATE). The recommendations and discussion below are based solely on those briefings; DHS's bioterrorism risk assessment was not made available to the committee in time for this interim report; however, the committee believes that these briefings included sufficient detail to adequately present the methodology used in the risk analysis.

NBACC has contracted with Battelle to produce a computational engine that assesses the "normalized risk" of 28 pathogens as that risk relates to death, morbidity, and direct economic costs.[5] In federal Fiscal Year 2007 (FY2007), DHS intends to improve and refine its probabilistic risk assessment (PRA). The committee has been asked to recommend possible directions for improvement, as well as to comment on the technical aspects of DHS's technique and the broader suitability of PRA. These comments are intended to provide guidance to DHS for its work during FY2007. Specifically, the committee has been given the following charge for this interim report:

- Assess the adequacy of the DHS's current methodology as a foundation for the desired risk analysis capabilities;
- Identify any other risk analyses that rely on the major components of the existing methodology, probabilistic risk analysis and multi-attribute risk analysis, and which could guide DHS's future developments;
- Assess the feasibility of incorporating models of second-order economic effects into the methodology during FY07;
- Identify better methods, if any, for handling the high degrees of uncertainty associated with the risk analyses of biological agents;
- Recommend near-term improvements to enhance the transparency of the method and its usefulness to decision makers;
- Discuss how the methodology could be extended to risks associated with classes of agents, including enhanced or engineered agents that have yet to be developed; and
- Discuss the feasibility of extending the methodology to also serve as a framework for risk analysis of chemical or radioactive threats.

For this interim report, the committee was not able to address the last of these tasks—to examine risk analysis for chemical or radioactive threats—because the breadth of this task exceeds the information that could be provided during briefings to the committee in one meeting. That task, however, will be addressed in the committee's final report.

The committee's charge for its final report is as follows:

- Recommend how the methodology can incorporate changing probability distributions that reflect how various actors (e.g., terrorists, first responders, public health community) adjust their choices over time or in different contexts;
- Recommend further improvements to the consequence analysis component of the methodology, including its models of economic effects;
- Identify any emerging methods for handling large degrees of uncertainty (e.g., fuzzy logic, possibility analysis) that merit consideration for future incorporation;

---

[5]In general usage, the distinction between "direct" and "indirect" costs is not precise. "Direct" refers to costs such as those associated with closing a facility or controlling an epidemic. Other, or "indirect," costs are those that result from these actions, such as lost business associated with the closing of a facility or reduced productivity due to public health measures.

- Recommend further improvements to the transparency and usability of the methodology;
- Discuss in more detail beyond the first report how the methodology could be extended to risks associated with classes of agents, including enhanced or engineered agents that have yet to be developed; and
- Discuss in more detail beyond the first report the feasibility of extending the methodology to also serve as a framework for risk analysis of chemical or radioactive threats.

This charge will require study of the issues addressed here in greater depth and with a view toward the longer term. The committee is confident, however, that the recommendations included in this interim report are appropriate and necessary in the near term. The committee's recommendations that follow address the general goal of improving methodology. Each recommendation relates to multiple elements of the charge, as noted in the accompanying text.

THE DHS BIOTERRORISM RISK ASSESSMENT

This interim report frequently refers to "risk" and activities surrounding its manipulation. For purposes of clarity, several definitions are given:

- *Risk*—the potential for realization of unwanted, adverse consequences to human life, health, property, or the environment, computed as the product of the probability of an event and the consequence of that event.
- *Risk analysis*—the overall process that involves risk assessment, risk perception, risk communication, and risk management. The hazards to be analyzed (e.g., physical, chemical, nuclear, radiological, and biological agents) may result from natural events (e.g., earthquakes and hurricanes), technological events (e.g., chemical accidents), and human activity (e.g., design and operation of engineered systems or attack by terrorists).
- *Risk assessment*—the scientific process of identifying hazards and quantifying their potential adverse consequences (magnitude, spatial scale, duration, and intensity) and associated probabilities including the uncertainties surrounding these estimates. Risk assessment may include a description of the cause-and-effect links among hazards and the nature of the interdependencies, vulnerabilities, and consequences.
- *Risk perception*—beliefs held by individuals or organizations about the risks of a hazard. Risk perception is concerned with psychological and emotional factors, which have been shown to have an enormous impact on behavior. Risk perception can be influenced by personal knowledge, experience, and beliefs; it can be affected by changing perceptions of the threat, the vulnerabilities, and/or the consequences; it may be influenced by information about hazards, risk assessments, risk policies, and risk management decisions.
- *Risk communication*—the process used by risk analysts, decision makers, policy makers, and intelligent adversaries to provide data, information, and knowledge to change the risk perceptions of individuals and organizations and enable them to assess the risk differently than they otherwise might. Risk communication needs must be considered when developing strategies for managing risk; thus any risk analysis methodology must take into account how affected individuals perceive and understand risk.
- *Risk management*—the process of constructing and evaluating strategies for reducing losses from future hazards and dealing with the recovery process should a disaster

occur. Risk management strategies include a combination of options, such as providing information (i.e., risk communication), economic incentives (e.g., subsidies, fines), insurance, compensation, regulations, and standards. These strategies enable individuals and private-sector or public-sector organizations to transfer, mitigate, or accept their perceived risks. Risk management strategies can be evaluated by undertaking cost-benefit analyses to determine the trade-off between the reduction of risk and the costs of undertaking such measures. In evaluating a risk management strategy, one needs to be concerned with the way resources are allocated (i.e., efficiency considerations) as well as with the impact of these measures on different stakeholders (i.e., distribution or equity considerations).

The model used for the DHS bioterrorism risk assessment is a computer-based tool used for assessing the relative risk of terrorist use of each of 28 specific pathogens, identified in other sources. The methodology described below is an instance of probabilistic risk assessment, which is particularly well adapted for low-frequency, high-potential-consequence events for which there is no database sufficient to assess risk using statistical analysis of historical data.

The PRA used by DHS divides the spectrum of possible attacks into a discrete set of scenarios, or sequences of events, and for each scenario it provides an estimate of the scenario's probability of occurrence, consequences, and risk. Owing to the extremely large size of the sample space, Battelle sampled the events in the scenarios involving a particular pathogen, estimated the risk associated with that pathogen, and compared it with the risk of other pathogens in order to obtain risk relative to that of other pathogens.

Each scenario involves a chain of as many as 17 events, which can be partitioned into those characterizing the terrorist group's motivations and goals; those involving its methods and ability to acquire, produce, and transport the given bioagent; and those surrounding the attack and response to it. Each event is further given discrete characteristics. For instance, the event of target selection can be further decomposed into the selection of a large, open building; a small enclosure; a large, divided building; a large outdoor space; a water pathway; a food pathway; or a contact target such as a letter. The event tree[6] generated thus has millions of scenarios, or paths through the tree, for which the probabilities and consequences must be explicitly or implicitly calculated.

For each scenario, a range of consequences—measured in terms of illnesses, fatalities, and economic losses—must be computed, with a probability distribution over the range. The "consequence engine" used for these computations consists of a series of equations whose variables are derived from the properties of the pathogen, the details of the scenario, and the hypothesized U.S. response to the terrorist event. DHS is developing improved means to estimate the first- and second-order economic effects (as discussed later in this report). In addition, it is developing systems dynamics models of the ways in which the scenarios might unfold. The committee will review this systems dynamics approach in its final report.

Even from this brief description, it can be seen that the DHS model requires a large amount of information, much of which is uncertain. This information includes the known properties of the pathogens, estimates of the propensities of terrorists to take different actions, and estimates of the reactions of the affected population and of the timeliness and effectiveness of the government response. With the exception of known scientific information, the parameters are either estimated from historical experience or elicited from experts, often in the form of probability distributions.

---

[6]An "event tree" is a visual representation of all events that can occur in a system. As the number of events increases, the picture fans out like the branches of a tree.

RECOMMENDATIONS

For the most part, the analysis described in the previous section follows approaches considered technically sound and useful in other areas of risk analysis such as nuclear reactor safety and chemical safety. In validation of risk, PRA avoids many of the practical problems and difficulties that arise from other alternative methods such as fuzzy logic, the analytic hierarchy process, or worst-case analysis (Banks and Anderson, 2006; Laviolette et al., 1995).

Event-tree analysis, which is the basis of PRA, is a well-developed risk tool in nuclear reactor safety and many other, usually engineering, contexts (Lindley and Singpurwalla, 1986). The main concern of the committee is that the current PRA event-tree paradigm does not fully support any of the components of risk analysis. It does not include consideration of the actions of an intelligent and reactive adversary, which is required for a complete risk analysis. It makes no provision for risk perception. It does not allow the exploration by decision makers of "what-if" questions, which is needed for risk management.[7] DHS needs to provide analyses for a variety of purposes to a variety of customers, and all within the context of competing security demands in the short run, while taking into account the longer-run concerns that may change over time. Therefore, a necessary first step is to clarify the longer-term goals and objectives of bioterrorism risk analysis.

Recommendation 1: DHS should establish a clear statement of the long-term purposes of its bioterrorism risk analysis.

In order to justify the current methodology as a foundation for future analyses, a clear statement of the long-term purposes of the bioterrorism risk analysis is needed to enunciate how it will support risk assessment, risk perception, and especially risk management decision making. Criteria and measures should be specified for measuring how well these purposes are achieved. Key issues to be addressed by such a statement should include the following: who the key stakeholders are; what their short- and long-term values, goals, and objectives are; how these values, goals, and objectives change over time; how the stakeholders perceive the risks; how they can communicate these risks more effectively; what they need from the risk assessment in order to make better (more effective, confident, rational, and defensible) resource-allocation decisions; and who should perform the analyses (contractors, government, both). Another important operational consideration is the determination of how DHS should incorporate new information in its analyses. The pace of change in biotechnology will require frequent and systematic updates of information used by the model. DHS issues "tailored assessments" to respond to unscheduled requirements, in addition to its biennial report, and it must be able to incorporate new intelligence information or technological change, for instance, in these analyses.

DHS's purposes for its bioterrorism risk assessment must be supported by its customers, by the U.S. Congress, and by the scientific community, among others; thus, DHS should actively solicit the opinions of its stakeholders to ensure that communication on issues of risk analysis is two-way. To that end, the language and analyses used must be precise. The technical presentations given to the committee suggest that the model documentation does not always use standard and consistent terminology. For example, several speakers at the committee's first meeting used the term "relative risk" to refer to what should be called "normalized risk," and "likelihood" was sometimes used as a synonym for "probability."

[7]The DHS methodology, as reflected in software, actually does allow changes in assumptions; but this must be done through an analyst and would require a significant time delay and limit the range of alternatives that could be examined.

The terms "risk," "expected risk," and "expected consequences" were often casually interchanged, and the computation of "normalized risk" was flawed.[8] The terms "illness" and "morbidity" should be clarified and defined more precisely (i.e., illness would need to be defined as either "infected" or "symptomatic").

Other terms used in the presentations to the committee were not precisely defined, and functional notation was confusing. DHS should define and use a standard lexicon, clarify concepts, and align with contemporary literature in order to improve the transparency of its models and results. DHS's operational definition of "risk" should be refined to include time explicitly—for example, by indicating how many events with various degrees of severity of adverse consequences can be expected over what time intervals if different risk management interventions are implemented. Attention also needs to be given to the uncertainty and ambiguity associated with these risks. Use of outside peer reviews may help in this regard. The issues raised here are not minor concerns; this lack of precision can lead to internal inconsistencies in the model and to communication problems at all levels.

DHS's risk assessment currently encompasses what are mainly traditional bioagents. However, it seems logical that the DHS vision for risk analysis should be broad enough to include risks posed by other significant future biological threats. Traditional bioagents are "naturally occurring microorganisms or toxin products with the potential to be weaponized and disseminated to cause mass casualties."[9] Testing the methodology by using existing biological agent threat lists, as has been done to date, is a prudent and logical way to start, given the very large number of pathogens that could possibly be used as weapons. Existing threat lists (e.g., from the Centers for Disease Control and Prevention[10]) reflect extensive experience and the judgment of the intelligence and scientific communities. However, many bioterrorism experts would agree that the "logic behind biowarfare programs of the past will not necessarily guide the life sciences as new technology rapidly emerges; biowarfare programs of the past predated current knowledge of molecular biology" (Relman, 2006, pp. 113-115). Therefore, future iterations of the methodology should also consider enhanced, emerging, and advanced agents in addition to traditional bioagents:

- *Enhanced agents* are those that are modified to circumvent current countermeasures—for example, microorganisms that are purposefully manipulated to be resistant to multiple antibiotics, thus complicating a public health response in the aftermath of an attack.
- *Emerging agents* are those that occur naturally but are newly recognized or anticipated to pose a public health threat—for example, a highly lethal and readily transmissible influenza strain that may cause a pandemic.
- *Advanced agents* are novel microorganisms that may be created by employing laboratory methods.

The results of such an extended risk assessment would be useful in determining the appropriate allocation of resources to develop flexible defenses—those that may be useful against a wide range of microorganisms that may share common processes in causing

---

[8]After normalization (division by the average risk over all agents), information about the actual magnitude of the risk is lost, affecting risk assessment and making the analysis of most resource-allocation decisions difficult. Moreover, distributions of risk, as normalized in this way, cannot be created by simply normalizing the scale of the non-normalized risk.

[9]*Federal Register*, Vol. 71, No. 174, 2006, available at http:/www.hhs.gov/ophep/ophemc/bioshield/PHEMCESStrategyFRNotice090806.pdf. Accessed Nov. 1, 2006.

[10]*Federal Register*, Vol. 71, No. 174, 2006, available at http://www.hhs.gov/ophep/ophemc/bioshield/PHEMCEStrategyFRNotice090806.pdf. Accessed Nov. 1, 2006.

disease. Such an assessment would require information that is not currently available—estimates of likely developments in biotechnology that would enable new capabilities that could be used by terrorists. The committee believes that, for the near term, the elicitation of expert opinion, similar to what was undertaken in DHS's assessment of traditional bioagents, would be a useful starting point. This could be the first step in establishing the risk imposed by agents not yet in the environment and in broadening the analysis to include classes of agents rather than individual agents. The committee will examine this difficult problem in more depth in its final report.

Recommendation 2: DHS should improve its analysis of intelligent adversaries.

Event trees were not originally developed to model intelligent adversaries who adapt their attacks in response to (or in anticipation of) their opponents' defensive actions and/or in response to their own initial successes or failures in plan execution. Alternative risk analysis techniques, including attack-preference, decision-tree, attack-tree, or attack-graph models,[11] can complement or replace probability elicitation. There have been recent advances in dealing with interdependent and coordinated adversary actions, called interdependent security (Heal and Kunreuther, 2005), which may improve the fidelity of DHS models.

To use a PRA event-tree risk assessment in the analysis of intelligent adversaries, the tree must include all realistic threats that adversaries may pursue. The committee believes that the DHS PRA tree is reasonably complete, although DHS should examine this further in light of the expectation that adversaries will adapt to any defensive decisions made by the United States. A small number of well-chosen red teams (i.e., individuals including both technologists and those with experience in targeting and strategy, whose purpose is to simulate adversarial decision making) to provide input for "what-if" scenarios can help to confirm and expand the current state of understanding and model validation and can complement expert opinion.

The probabilities in the event tree must be of sufficient quality to produce trustworthy results. Most of the event probabilities have been generated using expert opinion. DHS is keenly aware that this approach may be unreliable, and the committee is pleased that DHS intends to use CREATE's expertise to improve elicitation of the views of subject-matter experts. But the reliability of these probability assessments will always be problematic, requiring careful attention to the elicitation methods as well as needing well-designed sensitivity analyses (Kahneman and Tversky, 2000; Meyer and Booker, 2001). Moreover, strictly probabilistic analysis should also be supplemented with other methods, such as attack-preference models and attack-tree models, in order to ascertain any severe contradictions in the resulting risk management (or mitigation) recommendations.

The Mission Oriented Risk and Design Analysis (MORDA) model, used in several Department of Defense risk assessment studies, is an example of the use of subject-matter expert teams from various disciplines to collect data and incorporate expert knowledge about adversaries. The MORDA model uses this collected information in adversary models and attack-tree models (Buckshaw et al., 2005).

In order to better understand the sources of uncertainty and to plan for their reduction,

[11]Attack trees and attack graphs are modeling techniques for understanding risk in complex situations. Both are graphical representations showing all ways to attack or damage a system. Decision trees are event trees with decisions represented as possible events. Attack-preference models examine decisions from the viewpoint of the attacker rather than the defender. See http://csdl2.computer.org/persagen/DLAbsToc.jsp?resourcePath=/dl/proceedings/&toc=comp/proceedings/itcc/2004/2108/01/2108toc.xml&DOI=10.1109/ITCC.2004.1286496. Accessed Nov. 1, 2006.

any analysis resulting from the PRA model should include a data-quality matrix with a qualitative assessment of the sources and quality of the data and perhaps quantitative indications of the confidence and precision associated with current estimates (e.g., plausible range of values for model inputs) for the 28 bioagents and the 17 steps in the event tree developed by Battelle.

The committee believes that static probabilities, as they are currently used by DHS, are insufficient to model the behavior of intelligent adversaries. Static probabilities may be appropriate when dealing with nuclear reactors, but not for an intelligent adversary who adapts an attack on the basis of the actions of the defenders and on information that it acquires as planning and execution progress. Although classical game theory is a formal way to handle such situations, there is now a growing literature that may be more relevant for dealing with the adversarial nature of the bioterrorism problem (Bier et al., 2005; Enders and Sandler, 2006; Heal and Kunreuther, 2005). Studies have been conducted by the Navy Postgraduate School in which the defender computed a strategy that would minimize the maximum damage that could be caused by an attacker (terrorist) who was aware of that strategy. These "attacker-defender" studies, which have been undertaken in various contexts to determine how best to protect U.S. infrastructure, might serve to complement the static probability analyses currently used by DHS (Brown et al., in press).

Any analysis of adversarial actions, as well as of mitigation strategies and responses, will require accurate estimates of the real damages that the United States would experience. Currently, the PRA computes measures of mortality, morbidity, and direct economic costs. But indirect economic costs (e.g., of business interruption) must also be included to avoid underestimating true financial consequences. If these indirect costs are large, it may be necessary to evaluate their impact, taking into account risk aversion and/or loss aversion.[12]

Evaluation of these costs will require that DHS more carefully consider its consequence measures and modeling, which should be augmented to include indirect economic effects. DHS is planning to use input-output models and CREATE-developed general equilibrium models to improve its estimates of the direct economic consequences of terrorist events in its FY08 risk assessment. Both of these techniques can be used to estimate the indirect costs. The committee agrees that their use is appropriate for the next stage of model development.

DHS is planning, however, to pursue consequence modeling that is of higher fidelity and resolution than that of the modeling being used now. Such a path is not clearly justified by either data availability or currently articulated decision needs. More fine-grained and detailed consequence models of targets should only be pursued if such granularity directly supports improved risk management decision making. The committee is concerned about the use of too fine a granularity in the simulation. It could result in false precision that might be mistaken for accuracy in a model that is, by necessity, not particularly well validated, affecting both risk assessment and risk management. In addition, too fine a granularity decreases the transparency of the model. The committee is concerned that merely increasing the number of parameters that need to be elicited may not increase the real or useful precision of the model.

Individuals' perceptions of risks can have a major influence on indirect economic consequences, resulting in a need to develop strategies to manage risk perception and to deal with these perceptions. DHS should consider decision-analytic methods for dealing with

---

[12]Risk aversion is the reluctance of a person to accept a bargain with an uncertain payoff rather than another bargain with a more certain, but possibly lower, expected payoff. Loss aversion refers to the tendency for people to strongly prefer avoiding losses to acquiring gains.

issues such as attitudes toward probabilities and consequences (the components of risk), the role of affect and emotion, biases in judgment, and the types of rules used by individuals and groups in choosing between alternatives.

Recommendation 3: DHS should increase its risk analysis methodology's emphasis on risk management.

Risk managers should be able to explore the impact of different investment strategies on the effects they might have on the attacker. Typical trade-offs facing U.S. risk managers might involve allocating resources among human intelligence versus vaccine development or deployment of biohazard sensors. A given resource allocation may drive a corresponding set of decisions by potential terrorists, which in turn changes risks. The current DHS event-tree PRA is not adequate for such risk management purposes. This is so because the event-tree PRA cannot determine which portfolio of investments is most effective and how potential attackers are likely to respond, although it does provide value in giving a coarse look at relative risks. This inadequacy highlights the importance of improving the current risk analysis with red teaming, attack-preference models, attack-tree models, and perhaps, game-theoretic analyses or alternatives. All of these techniques will serve to mitigate the high degree of uncertainty associated with the risk analysis of biological agents.

It is unclear to the committee how the current PRA approach supports DHS's design and evaluation of alternative risk management strategies. The computational engine does not permit, let alone encourage, risk managers to explore scenarios of "if resource allocation, then probable consequence." DHS needs to determine how alternative risk management strategies, involving specific resource investments in attack prevention, consequence mitigation, or other forms of protection, translate to changes in the overall level of risk. An interface and visualization component is needed to display results and limitations of this very complex model and to improve transparency.

In evaluating alternative risk management strategies, DHS should take into account all significant benefits that result from any strategy, beyond just those benefits that directly impact the risks of bioterrorism attacks. For instance, investment in intelligence might include all homeland security risks, and the risk management trade-offs should be considered in that larger context. This last conclusion has ramifications for all of DHS's risk analysis and directly addresses the committee's final charge. It will be more fully explored in this study's final report.

DHS should develop a targeted research program to develop risk analysis methods that take into account the decision maker's risk perception and risk management strategies. Such a program would include the following, for example: consideration of how constraints on resources available to the decision maker might affect terrorist decisions, and an understanding of how attackers who encounter failures or setbacks in executing an initial plan will respond—including the realistic possibility that they will implement contingency plans or adaptively replan to achieve goals that still appear feasible and worthwhile.[13] Methods for modeling multiple coordinated attacks by teams of adversaries should also be considered.[14] These changes should all be incorporated into the next generation of DHS's bioterrorism risk assessment and management technologies. The committee believes that these extensions can be achieved by expanding the models rather than by increasing the fidelity of existing models.

[13]See http://handle.dtic.mil/100.2/ADA009141. Accessed Nov. 1, 2006.

[14]See http://www.rms.com/Publications?QuanTerRisk4Portfolios_Woo_Aon.pdf. Accessed Nov. 1, 2006.

SUMMARY

As previously noted, each of the committee's recommendations relates to multiple elements of its charge. Here, responses to each element of the charge, in order, are summarized.

- DHS's current methodology is adequate but incomplete. A statement of purpose is needed, as well as methods to handle intelligent adversaries. Red teaming, attack-preference models, attack-tree models, and game-theoretic analyses should all be examined for the purpose of supplementing the existing methodology.
- The analyses cited, by Buckshaw et al. (2005) and by Brown et al. (in press), are examples of other types of risk analysis that would be appropriate for DHS's future development.
- DHS's current plans for the incorporation of second-order indirect economic effects into its methodology are appropriate, as long as the model's level of granularity is carefully considered.
- High degrees of uncertainty can be addressed by the incorporation of red teaming, attack-preference models, attack-tree models, and game-theoretic analyses. The incorporation of data-quality matrices in DHS's analyses will lead to a better understanding of the sources of uncertainty.
- In order to improve transparency, DHS should define and use a standard lexicon, clarify concepts, and align with the contemporary literature.
- In order to extend the methodology to risks associated with classes of agents, careful elicitation of expert opinion is the best starting point. This issue will be further examined in the committee's final report.
- No examination was made in this interim report of the feasibility of extending the methodology to serve as a framework for risk analysis of chemical or radioactive threats.

REFERENCES

Banks, D., and S. Anderson. 2006. "Combining Game Theory and Risk Analysis in Counterterrorism: A Smallpox Example." Pp. 9-12 in A. Wilson, G. Wilson, and D. Olwell, eds., *Statistical Methods in Counterterrorism*. New York: Springer.

Bier, Vicki, Santiago Oliveros, and Larry Samuelson. 2005. "Choosing What to Protect: Strategic Defense Allocation Against an Unknown Attacker." University of Wisconsin Working Paper.

Brown, G., W. Matthew Carlyle, Javier Salmeron, and Kevin Wood. In press. "Defending Critical Infrastructure." *Interfaces*.

Buckshaw, Donald L., Gregory S. Parnell, Willard L. Unkenhotz, Donald L. Parks, James M. Wallner, and O. Sami Saydjari. 2005. "Mission Oriented Risk and Design Analysis of Critical Information Systems." *Military Operations Research* 10(2): 19-38.

Enders, Walter, and Todd Sandler. 2006. *The Political Economy of Terrorism*. Cambridge: Cambridge University Press.

Heal, Geoffrey, and Howard Kunreuther. 2005. "You Only Die Once: Interdependent Security in an Uncertain World." Pp. 35-36 in H.W. Richardson, P. Gordon, and J.E. Moore II, eds., *The Economic Impacts of Terrorist Attacks*. Cheltenham, U.K.: Edward Elgar.

Kahneman, Daniel, and Amos Tversky. 2000. *Choices, Values and Frames*. New York: Cambridge University Press.

Laviolette, Michael, John W. Seamon, Jr., J. Douglas Barrett, and William H. Woodall. 1995. "A Probabilistic and Statistical View of Fuzzy Methods." *Technometrics* 37: 249-261.

Lindley, Dennis V., and Nozer D. Singpurwalla. 1986. "Reliability and Fault Tree Analysis Using Expert Opinions." *Journal of the American Statistical Association* 81: 87-90.

Meyer, M.A., and J.M. Booker. 2001. *Eliciting and Analyzing Expert Judgment: A Practical Guide*. ASA-SIAM Series on Statistics and Applied Probability, Vol. 7. Philadelphia, Pa.: Society for Industrial and Applied Mathematics.

Relman, D.A. 2006. "Bioterrorism—Preparing to Fight the Next War." *New England Journal of Medicine* 354: 113-115.

Appendix

AGENDA FOR COMMITTEE MEETING, AUGUST 28-29, 2006

KECK CENTER OF THE NATIONAL ACADEMIES

500 Fifth Street, N.W.
Washington, DC 20001

Monday, August 28, 2006

Closed Session (committee members and NRC staff only)

8:00 a.m.

Data-Gathering Session Open to the Public

9:45 a.m.	Introductory Remarks	Department of Homeland Security Science and Technology Leadership
10:00 a.m.	Biology Presentation (background for non-biologists)	Prof. Luciana Borio, University of Pittsburgh, Center for Biosecurity
10:45 a.m.	Break	
11:00 a.m.	DHS and National Biodefense Analysis and Countermeasures Center (NBACC) Background and Risk Assessment Requirements	Dr. Steven Bennett, DHS/NBACC Dr. Bernard Courtney, DHS/NBACC

11:30 a.m.	DHS 2006 Bioterrorism Risk Assessment Methodology	Dr. Richard Denning, Battelle Memorial Institute
1:00 p.m.	Lunch	
1:45 p.m.	Past Experiences and Implications for Bioterrorism	Prof. Detlof von Winterfeldt, Director, Center for Risk and Economic Analysis of Terrorism Events (CREATE), University of Southern California
2:15 p.m.	Assessing the Economic Impacts of Terrorism—Capturing Behavioral Linkages and Resilience	Prof. Adam Rose, Pennsylvania State University and CREATE
2:45 p.m.	Break	

Data-Gathering Session Open to the Public: Scenario Analysis and Consequence Modeling

3:00 p.m.	Branch Probabilities and Uncertainty Management	Mr. Rob Carnell, Battelle
	Atmospheric (Outdoor) Dispersion Modeling	Ms. Mary Shell, Battelle
	Indoor Aerosol Dispersion Modeling	Dr. Brian Hawkins, Battelle
	Medical Mitigation and Epidemiological Modeling	Ms. Traci Hale and Dr. Nancy McMillan, Battelle
	Food and Water Contamination Modeling	Mr. Jon David Sears, Battelle
	Risk Calculation Engine	Mr. Rob Carnell, Battelle
5:30 p.m.	Reception	

Tuesday, August 29, 2006

Data-Gathering Session Open to the Public

9:30 a.m.	Updates and Planned Changes for the 2008 Bioterrorism Risk Assessment	DHS/NBACC, Battelle Staff
10:45 a.m	Break	

Closed Session (committee members and NRC staff only)

4:00 p.m.	Adjourn

Appendix K

Meeting Agendas

AUGUST 28-29, 2006

Monday, August 28, 2006

Closed Session (Committee Members and NRC Staff Only)

8:00 a.m.

Data-Gathering Session Open to the Public

9:45 a.m.	Introductory Remarks	Department of Homeland Security (DHS) Science and Technology Leadership
10:00	Biology Presentation (Background for Nonbiologists)	Prof. Luciana Borio, Center for Biosecurity, University of Pittsburgh
10:45	Break	
11:00	DHS and National Biodefense Analysis and Countermeasures Center (NBACC) Background and Risk Assessment Requirements	Dr. Steve Bennett, DHS/NBACC Dr. Bernard Courtney, DHS/NBACC
11:30	DHS 2006 Bioterrorism Risk Assessment Methodology	Dr. Richard Denning, Battelle Memorial Institute
1:00 p.m.	Lunch	
1:45	Past Experiences and Implications for Bioterrorism	Prof. Detlof von Winterfeldt, Director, Center for Risk and Economic Analysis of Terrorism Events (CREATE), University of Southern California
2:15	Assessing the Economic Impacts of Terrorism—Capturing Behavioral Linkages and Resilience	Prof. Adam Rose, Pennsylvania State University and CREATE
2:45	Break	

NOTE: Meetings of the Committee on Methodological Improvements to the Department of Homeland Security's Biological Agent Risk Analysis were held at the Keck Center of the National Academies, 500 Fifth Street, N.W., Washington, DC 20001.

3:00	Branch Probabilities and Uncertainty Management	Mr. Rob Carnell, Battelle
	Atmospheric (Outdoor) Dispersion Modeling	Ms. Mary Shell, Battelle
	Indoor Aerosol Dispersion Modeling	Dr. Brian Hawkins, Battelle
	Medical Mitigation and Epidemiological Modeling	Ms. Traci Hale and Dr. Nancy McMillan, Battelle
	Food and Water Contamination Modeling	Mr. Jon David Sears, Battelle
	Risk Calculation Engine	Mr. Rob Carnell, Battelle
5:30 p.m.	Reception	

Tuesday, August 29, 2006

Data-Gathering Session Open to the Public

9:30 a.m.	Updates and Planned Changes for the 2008 Bioterrorism Risk Assessment	DHS/NBACC, Battelle Staff
10:45	Break	

Closed Session (Committee Members and NRC Staff Only)

11:00 a.m.		
4:00 p.m.	Adjourn	

NOVEMBER 19-20, 2006

Sunday, November 19, 2006

Closed Session (Committee Members and NRC Staff Only)

8:00 a.m.

Data-Gathering Session Open to the Public

10:30 a.m.	Break	
11:00	Manufactured Bioagents	Prof. Stephen Morse, Director, Center for Public Health Preparedness at the Mailman School of Public Health, Columbia University
12:00 noon	Lunch	
1:00 p.m.	Emerging Methods for Handling Large Degrees of Uncertainty	Dr. Alyson Wilson, Technical Staff Member, Statistician and Technical Lead, Department of Defense Programs, Los Alamos National Laboratory
2:00	Strategies for Adversarial Risk Analysis	Prof. David Banks, Institute of Statistics and Decision Sciences, Duke University
2:30	Frequentist Approach to Risk Analysis	Prof. Tapan Nayak, Department of Statistics, George Washington University
3:30	Break	

Closed Session (Committee Members and NRC Staff Only)

3:45 p.m.

5:30 p.m.	Reception	

Monday, November 20, 2006

Closed Session (Committee Members and NRC Staff Only)

8:00 a.m.

Data-Gathering Session Open to the Public

9:30 a.m.	DHS Chemical Agent Risk Analysis	Dr. George Famini, DHS

Closed Session (Committee Members and NRC Staff Only)

10:30 p.m.

FEBRUARY 9-10, 2007

Friday, February 9, 2007

Closed Session (Committee Members and NRC Staff Only)
8:00 a.m.

Data-Gathering Session Open to the Public[1]

8:45 a.m.	Medical Response and Preparedness for a Radiological/Nuclear Event	Dr. Norman Coleman, National Institutes of Health Dr. Peter Highnam, Public Health Emergency Medical Countermeasures (PHEMC)/Assistant Secretary for Preparedness and Response (ASPR)/Department of Health and Human Services (HHS)
9:45	Perspectives on Risk Assessment for a Global Nuclear Detection Architecture	Mr. Mark Mullen, Lead Systems Architect, Defense Nuclear Detection Office/DHS
10:30	Break	
10:45	Strategic Biodefense	Prof. Tara O'Toole, University of Pittsburgh
11:45	Lunch	
12:30 p.m.	Systems Dynamics Approach to the Spread of Infectious Disease	Ms. Cheryl Dingus, Battelle Ms. Michelle Gisi, Battelle
1:30	The Spread of Infectious Disease	Prof. Marc Lipsitch, Harvard University

[1]The committee deviated from this published schedule to hear an open briefing from Rear Admiral Jay Cohen, Undersecretary of Science and Technology of the Department of Homeland Security: "DHS Science and Technology: Enabling Technology to Protect the Nation," from approximately 11:30 to 12:30.

Saturday, February 10, 2007

Open Session[2]

8:00 a.m.	DHS Reaction to Interim Report Changes at DHS	Dr. Steve Bennett, DHS
10:30	Break	
10:45	Institute for Defense Analyses Approach to Risk Assessment for Critical Infrastructure	Dr. James Morgensen, IDA
11:45	Lunch	

Closed Session (Committee Members and NRC Staff Only)

12:30 p.m.

MAY 18-19, 2007

Friday, May 18, 2007

Closed Session (Committee Members and NRC Staff Only)

8:00 a.m.

Saturday, May 19, 2007

Closed Session (Committee Members and NRC Staff Only)

8:00 a.m.

---

[2]This included the briefing: "2008 DHS Bioterrorism Risk Assessment: Planned Improvements," by Traci Hale of Battelle Memorial Institute.

Appendix L

Biographies of Committee Members

Gregory S. Parnell, *Chair*, is professor of systems engineering at the United States Military Academy at West Point and teaches decision and risk analysis, systems engineering, and operations research. His research focuses on decision analysis, risk analysis, resource allocation, and systems engineering for defense, intelligence, homeland security, research and development (R&D), and environmental applications. He co-edited *Decision Making for Systems Engineering and Management, Wiley Series in Systems Engineering* (Wiley and Sons, 2008), and has published more than 100 papers and book chapters. He is a member of the Chief Technology Officer and Information Assurance Panels of the National Security Agency Advisory Board and is a former member of the Department of Energy's Environmental Management National Prioritization Team. He is a senior principal with Innovative Decisions, Inc., a decision and risk analysis firm, and a former principal with Toffler Associates, a strategic advisory firm. Dr. Parnell is a former president of the Decision Analysis Society of the Institute for Operations Research and Management Science (INFORMS) and of the Military Operations Research Society (MORS). He has also served as editor of *Journal of Military Operations Research*. Dr. Parnell is a retired Air Force colonel with experience in space operations, R&D management, and operations research. Dr. Parnell received his Ph.D. from Stanford University and is a graduate of the Industrial College of the Armed Forces. He has received several professional awards, including the United States Army Dr. Wilbur B. Payne Memorial Award for Excellence in Analysis, MORS Clayton Thomas Laureate, two INFORMS Koopman Prizes, and the MORS Rist Prize. He was elected a fellow of the MORS in 1997 for his contributions to military operations research.

David Banks is a professor in the Department of Statistical Science at Duke University. He is currently chair of the American Statistical Association (ASA) Section on Statistics in Defense and National Security and is a past chair of the Section on Risk Analysis. He is editor of the *Journal of the American Statistical Association*, a member of the board of directors of the ASA, and a former member of the ASA's Committee on Applied and Theoretical Statistics.

Luciana L. Borio, M.D., is senior associate at the Center for Biosecurity of the University of Pittsburgh Medical Center and assistant professor of medicine at the University of Pittsburgh. She also serves part time at the U.S. Department of Health and Human Services (HHS) as an adviser on biodefense programs. She is an infectious disease physician and continues to practice medicine at Johns Hopkins Hospital. Dr. Borio's work focuses on policies to improve the nation's preparedness for bioterrorism, by supporting threat assessments, medical countermeasures development, and medical response plans. Dr. Borio is an associate editor of the peer-reviewed journal *Biosecurity and Bioterrorism: Biodefense Strategy, Practice, and Science*, and she is co-managing editor of the Clinicians' Biosecurity Network, a real-time, online communications network designed to facilitate communications among physicians during health care crises. She serves on the Global and Public Health Committee and the Bioemergencies Task Force of the Infectious Diseases Society of America. She has lectured extensively and has published a series of manuscripts and book chapters on biodefense-related issues. Dr. Borio is a member of the Infectious Diseases Society of America, Phi Beta Kappa, and Alpha Omega Alpha. Prior to joining the Center for Biosecurity at its founding in 2003, she was a senior fellow at the Johns Hopkins University Center for Civilian Biodefense Strategies and assistant professor of medicine in the Division of Infectious Diseases at Johns Hopkins University. In 2002, Dr. Borio left the Johns Hopkins Center to work full time as senior health advisor at HHS. There she implemented and managed mathematical modeling projects to assess the health effects of bioterrorism on civilians and to inform medical countermeasures procurement activities for the Office of Preparedness and Response. She rejoined the Johns Hopkins Center in 2003 and continues to serve part time at HHS, where she advises on the requirements for and development of medical countermeasures. She received a B.S. in 1992 and an M.D. in 1996 from the George Washington University. She completed residency in 1999 in internal medicine at the New York Presbyterian Hospital-Cornell Medical Center, and subsequently completed a combined fellowship in infectious diseases (at Johns Hopkins University) and critical care medicine (at the National Institutes of Health).

Gerald G. Brown is Distinguished Professor of Operations Research at the Naval Postgraduate School, where he has taught and conducted basic and applied research in optimization theory and optimization-based decision support since 1973, earning awards for both outstanding teaching and research. His military research has been applied by every uniformed service, in areas ranging from strategic nuclear targeting to capital planning. Professor Brown has been awarded the Rist Prize for military operations research and has been credited with guiding investments of more than a trillion dollars. He has designed and implemented decision support software currently used by two-thirds of the Fortune 50 companies, in areas ranging from vehicle routing to supply-chain optimization. His research appears in scores of open-literature publications and classified reports, many of which are seminal references in the field. He is also a fellow of the Institute for Operations Research and Management Science and is a founding director of Insight, Inc., the leading provider of strategic supply-chain optimization-based decision support tools to the private sector. He is a retired naval officer and was recently elected to the National Academy of Engineering.

Anthony Cox, Jr., is president of Cox Associates, an independent, Denver-based applied research and consulting company specializing in wireless and optical network design and optimization software tools, customer data mining and predictive modeling, and decision and risk analysis technologies. Dr. Cox has a Ph.D. in risk analysis and an S.M. in operations research, both from the Massachusetts Institute of Technology's Department of Electrical Engineering and Computer Science; and an A.B. from Harvard University. Prior to starting Cox Associates in 1986, he consulted in risk analysis, economics and statistics, operations research, and artificial intelligence at Arthur D. Little, Inc., in Cambridge, Massachusetts. From 1987 to 1996, he managed applied research and high-technology product development efforts for US WEST Advanced Technologies in Boulder, Colorado. He was senior director of advanced communications research, business and engineering modeling, and network architectures. He is currently an honorary full professor of mathematics at the University of Colorado at Denver, where he lectures on topics in biomathematics, health risk modeling, computational statistics, and machine learning. Dr. Cox is on the faculties of the Center for Computational Mathematics and the Center for Computational Biology at the University of Colorado at Denver and is clinical professor of preventive medicine and biometrics at the University of Colorado Health Sciences Center, where he teaches and guides graduate research on uncertainty analysis and causation in epidemiological studies. He is on the editorial board of *Risk Analysis: An International Journal* and is co-editor of the *Journal of Heuristics*. He is a full member of the Institute for Operations Research and the Management Sciences, the Society for Risk Analysis, and the American Statistical Association. He has chaired numerous conference sessions on various aspects of risk, uncertainty, network design, and optimization. Dr. Cox was elected to the New York Academy of Sciences in 1992 and was made a lifetime fellow of the Society for Risk Analysis in 1993. In 1994, he was a recipient of the Operations Research Society of America's prestigious ORSA prize for the best real-world applications of operations research having profound business impact. In addition to hands-on experience and professional activities in telecommunications decision and risk analysis, operations research, artificial intelligence, and applied statistics, Dr. Cox has authored and co-authored more than 100 journal articles and book chapters on advanced aspects of these fields. He holds more than a dozen U.S. and international patents on applications of network optimization, speech recognition, and signal processing technologies in telecommunications.

John Gannon is vice president for global analysis at BAE Systems. He joined BAE Systems after serving as staff director of the U.S. House of Representatives Homeland Security Committee, the first new committee established by Congress in more than 30 years. In 2002-2003, he was a team leader in the White House's Transitional Planning Office for the Department of Homeland Security. He served previously in the senior-most analytic positions in the intelligence community, including as the Central Intelligence Agency's director of European analysis, deputy director for intelligence, chairman of the National Intelligence Council, and assistant director of central intelligence for analysis and production. In the private sector, he developed the analytic workforce for Intellibridge Corporation, a Web-based provider of outsourced analysis for government and corporate clients. He served as a naval officer in Southeast Asia and later in several Naval Reserve commands, retiring as a captain. Dr. Gannon has a bachelor's degree from Holy Cross College in Worcester, Massachusetts, and master's and doctorate degrees from Washington University in St. Louis. He is an adjunct professor in the National Security Studies Program at Georgetown University.

Eric Harvill is an associate professor of microbiology and infectious disease at Pennsylvania State University. After graduate studies in molecular immunology and postdoctoral research in bacterial pathogenesis, he established a group that examines the interactions between bacterial pathogens and the host immune system to determine the molecular bases for these complex interactions. More recently, Dr. Harvill has examined the evolution of closely related respiratory pathogens of the genus *Bordetella*, examining the genomic and genetic differences that distinguish persistent commensals of all the animals around us from the acute and virulent forms that infect nearly all humans, causing whooping cough only in those who are not vaccinated. His laboratory uses a combination of the approaches common to bacterial pathogenesis, bacterial genomics/transcriptomics, comparative biology, and molecular immunology to understand the evolution of these pathogens.

Howard Kunreuther is the Cecilia Yen Koo Professor of Decision Sciences and Public Policy at the Wharton School, University of Pennsylvania, as well as co-director of the Wharton Risk Management and Decision Processes Center. He has a long-standing interest in ways that society can better manage low-probability–high-consequence events as they relate to technological and natural hazards and has published extensively on the topic. He is a fellow of the American Association for the Advancement of Science and Distinguished Fellow of the Society for Risk Analysis, receiving the society's Distinguished Achievement Award in 2001. Professor Kunreuther has written or co-edited a number of books and papers, including *Catastrophe Modeling: A New Approach to Managing Risk* (with Patricia Grossi) and *Wharton on Making Decisions* (with Stephen Hoch). He is a recipient of the Elizur Wright Award for the publication that makes the most significant contribution to the literature of insurance.

Stephen S. Morse is founding director of the Center for Public Health Preparedness at the Mailman School of Public Health of Columbia University and is a full professor in the Epidemiology Department. He also holds an adjunct faculty appointment at the Rockefeller University. Dr. Morse returned to Columbia University in 2000 after 4 years in government service as program manager at the Defense Advanced Research Projects Agency of the Department of Defense. In that position, he co-directed the Pathogen Countermeasures program and subsequently directed the Advanced Diagnostics program. Dr. Morse was chair and principal organizer of the 1989 National Institute of Allergy and Infectious Diseases/National Institutes of Health Conference on Emerging Viruses and has served as an adviser to the World Health Organization, the Pan-American Health Organization, the Centers for Disease Control and Prevention, the Food and Drug Administration, and other agencies. He was the founding chair of ProMED (the nonprofit international Program to Monitor Emerging Diseases) and was one of the originators of ProMED-mail, a network inaugurated by ProMED in 1994 for outbreak reporting and disease monitoring using the Internet. Dr. Morse currently serves on the steering committee of the Institute of Medicine's (IOM's) Forum on Emerging Infections and was previously a member of other IOM committees. He is a fellow of the American Academy of Microbiology and the American College of Epidemiology, a life member of the Council on Foreign Relations, and serves on the National Research Council's standing Committee on Biodefense Analysis and Countermeasures. Dr. Morse received his Ph.D. from the University of Wisconsin-Madison.

Marguerite Pappaioanou is executive director of the Association of American Veterinary Medical Colleges (AAVMC). Before joining AAVMC on November 1, 2007, she had served the previous 3 years as professor of infectious disease epidemiology in the School of Public Health at the University of Minnesota, which followed a 21½ year career at the Centers for Disease Control and Prevention. Her areas of interests include emerging zoonotic infectious diseases, with a special interest in influenza viruses, malaria, and HIV; bioterrorism and agroterrorism; disease surveillance; and disease prevention and control. She actively promotes linking human and animal health and the use of data in formulating evidence-based health policies.

Stephen Pollock is Herrick Emeritus Professor of Manufacturing and emeritus professor of industrial and operations engineering at the University of Michigan. He has taught courses in decision analysis, mathematical modeling, dynamic programming, and stochastic processes. His recent research activities include developing cost-optimal monitoring and maintenance policies, sequential hypothesis testing, modeling large multiserver systems, and dynamic optimization of radiation treatment plans. Dr. Pollock was the director of the Program in Financial Engineering and the Engineering Global Leadership honors program. He has been area editor of *Operations Research,* senior editor of *IIE Transactions,* president of the Operations Research Society of America, and a senior fellow of The University of Michigan Society of Fellows. He is a founding fellow of the Institute for Operations Research and the Management Sciences, and was awarded its Kimball Medal in 2002. He was a member of the Army Science Board and is a member of the National Academy of Engineering.

Nozer D. Singpurwalla is professor of statistics and Distinguished Research Professor at the George Washington University in Washington, D.C. He has been a visiting professor at Carnegie Mellon University, Stanford University, the University of Florida at Tallahassee, and the University of California, Berkeley. During the fall of 1991, he was the first C.C. Garvin Visiting Endowed Professor in the Mathematical Sciences at the Virginia Polytechnic Institute and State University. He is fellow of the Institute of Mathematical Statistics, the American Statistical Association (ASA), and the American Association for the Advancement of Science, and he is an elected member of the International Statistical Institute. Dr. Singpurwalla is the 1984 recipient of the U.S. Army's S.S. Wilks Award for Contributions to Statistical

Methodologies in Army Research, Development and Testing and was the first recipient of The George Washington University's Oscar and Shoshana Trachtenberg Prize for Faculty Scholarship. He co-authored a standard book in reliability and has published 157 papers on reliability theory, warranties, failure data analysis, Bayesian statistical inference, dynamic models and time series analysis, quality control, and statistical aspects of software engineering. In 1993 he was selected by the National Science Foundation (NSF), ASA, and the National Institute of Standards and Technology (NIST) as the ASA/NIST/NSF Senior Research Fellow. In 1993 he was awarded a Rockefeller Foundation grant as a scholar in residence at the Bellagio, Italy, Center.

Alyson Wilson is a project leader, technical staff member, and the technical lead for Department of Defense programs in the Statistical Sciences Group at the Los Alamos National Laboratory. Prior to her move to Los Alamos, Dr. Wilson was a senior operations research systems analyst working in support of the U.S. Army Operational Evaluation Command, Air Defense Artillery Evaluation Directorate. She also spent 2 years at the National Institutes of Health performing research in the biomedical sciences. Her research focuses on Bayesian methods, with emphasis on reliability modeling and information combination. She is the past chair of the American Statistical Association Section on Statistics in Defense and National Security and chair of the American Statistical Association's President's Task Force in Defense and Security. She received her Ph.D. in statistics from the Institute of Statistics and Decision Sciences at Duke University.

Appendix M

Acronyms

9/11	September 11, 2001
BDM	Bioterrorist Decision Model
BTCC	Biological Threat Characterization Center
BTRA	Biological Threat Risk Assessment
CBA	cost-benefit analysis
CBRN	chemical, biological, radiological, nuclear
CDC	Centers for Disease Control and Prevention
CREATE	Center for Risk and Economic Analysis of Terrorism Events
DALY	disability-adjusted life-year
DHS	Department of Homeland Security
DNA	deoxyribonucleic acid
DSS	decision support system
EP	exceedance probability
EPA	Environmental Protection Agency
FY	fiscal year
GAO	General Accounting Office; *now* Government Accountability Office
HSPD	Homeland Security Presidential Directive
IDS	interdependent security
IL-4	interleukin-4
LHS	Latin Hypercube Sampling
NBACC	National Biodefense Analysis and Countermeasures Center
NRC	National Research Council
OMB	Office of Management and Budget
PCR	polymerase chain reaction
PDF	probability density function
PRA	probabilistic risk assessment
QALY	quality-adjusted life-year
RNA	ribonucleic acid
SARS	severe acute respiratory syndrome
SEIR	susceptible, exposed, infected, and recovered
SME	subject-matter expert
SRA	Society for Risk Analysis
TOPOFF	Top Officials
U.S. NRC	U.S. Nuclear Regulatory Commission
WMD	weapons of mass destruction